Advances in Optics and Optoelectronics

Series Editor

Perry Ping Shum, Southern University of Science and Technology, Shenzhen, China

The Advances in Optics and Optoelectronics series focuses on the exciting new developments in the fast emerging fields of optics and optoelectronics. The volumes cover key theories, basic implementation methods, and practical applications in but not limited to the following subject areas:

AI Photonics
Laser Science and Technology
Quantum Optics and Information
Optoelectronic Devices and Applications
Fiber-Based Technologies and Applications
Near-infrared, Mid-infrared and Far-infrared Technologies and Applications
Biophotonics and Medical Optics
Optical Materials, Characterization Methods and Techniques
Spectroscopy Science and Applications
Microscopy and Adaptive Optics
Microwave Photonics and Wireless Convergence
Optical Communications and Networks

Within the scopes of the series are monographs, edited volumes and conference proceedings. We expect that scientists, engineers, and graduate students will find the books in this series useful in their research and development, teaching and studies.

Lei Wei
Editor

Advanced Optical and Optoelectronic Fibers

Springer

Editor
Lei Wei
Nanyang Technological University
Singapore, Singapore

ISSN 2731-6009 ISSN 2731-6017 (electronic)
Advances in Optics and Optoelectronics
ISBN 978-981-97-6217-0 ISBN 978-981-97-6218-7 (eBook)
https://doi.org/10.1007/978-981-97-6218-7

This Springer imprint is published by the registered company Springer Nature Singapore Pte Ltd.
The registered company address is: 152 Beach Road, #21-01/04 Gateway East, Singapore 189721, Singapore

Preface

Advanced optical and optoelectronic fibers represent an interdisciplinary convergence of optical waveguides, optoelectronics, material engineering, micro/nanofabrication, biomedical interfaces, and wearable technology. This book emphasizes recent scientific and technological breakthroughs in various optical and optoelectronic fibers, highlighting emerging fiber materials, innovative in-fiber structures, and novel functionalities. The book is organized into nine chapters.

- Chapter 1 "Hollow-Core Antiresonant Fibers" focuses on the development history, waveguide principle, loss mechanism, and advanced applications of hollow-core antiresonant fibers (HC-ARFs). The fundamental waveguide principle of HC-ARFs is introduced, where the effects of antiresonant reflection and core-cladding mode phase mismatching are discussed, enabling advanced applications including data transmission, supercontinuum generation, optical frequency comb distribution, and gas detection.
- Chapter 2 "Specialty Fiber Technology for Brillouin Optical Time Domain Analysis" introduces the fundamentals of utilization of unique properties of specialty fibers in the Brillouin optical time domain analysis (BOTDA) system. This leads to enhanced sensor performance, including large-scale strain sensing, temperature-insensitive strain sensing, and high-temperature strain sensing, especially for curvature sensing, shape reconstruction, and discriminative sensing of temperature and strain.
- Chapter 3 "In-Fiber Mach–Zehnder Interferometers for Sensing" presents the basic principle, fabrication methods, and sensing mechanism of the in-fiber Mach-Zehnder interferometers (MZIs). They are applied for the measurement of temperature, strain, curvature, and refractive index as well as other measurands such as humidity, magnetic field, and the concentration of gas and metal ions.
- Chapter 4 "Beyond the Spectrum: Specialty Optical Fibers in Magnetic Field Sensing" explores the advancements in magnetic field sensors based on optical fibers and advanced multi-material fibers, demonstrating diverse applications such as high-sensitivity magnetic field imaging, environmental monitoring for geophysical exploration, and industrial applications in equipment monitoring.

The evolving landscape of fiber-shaped magnetic field sensors promises enhanced capabilities and broader applications across industries.

- Chapter 5 "Thermally Drawn Optoelectronic and Optofluidic Polymer Fibres for Brain Applications" discusses material selections, fabrication, and optimization to achieve optoelectronic and optofluidic polymer fibers for modulating and monitoring neural activity. This leads to the development of soft, biocompatible optical fibers able to deliver and collect light in deep brain regions over long periods with minimal inflammation.
- Chapter 6 "Advanced Optical and Optoelectronic Fibers for Bio-integrated Systems" presents optical and optoelectronic fiber-based devices, circuits, and systems in bio-integrated applications. With advancements in material design and mechanics, specific applications in detecting biophysical signals, biosensors, human-machine interfaces, and photo-biomodulation therapy are achieved.
- Chapter 7 "Fiber X-Ray Detectors" covers the development of fiber X-ray detectors with unique superiority in miniaturization, flexibility, and remote measurement, as well as enhanced radiative absorption and spatial resolution. Both detection mechanisms and preparation methods of fiber detectors are summarized, offering great potential in many fields, like scientific research, medical imaging, and nondestructive testing.
- Chapter 8 "Fiber Nanocomposites Based on Two-Dimensional Nanosheets" reports recent advancements in the preparation approaches, interface construct, mechanical and electrical properties, and applications of fiber nanocomposites, emphasizing their transformative impact on electrical functionalities for flexible electronics. The incorporation of diverse 2D nanosheets into fibrous structures leads to a paradigm shift in material engineering through interfacial interactions.
- Chapter 9 "Fiber-Based Wearable Sensors with Machine Learning" offers a comprehensive review of wearable devices that utilize both fiber sensors and machine learning. Various fiber sensors are discussed based on their working principles and dimensions, and an overview of the machine learning algorithms applied to these fiber sensors is presented. The advantages, limitations, and concerns of different approaches are highlighted, paving the way for the future development of intelligent wearable devices and systems.

I would like to thank all the chapter authors for their outstanding contributions. Moreover, I would like to thank Nanyang Technological University for the constant support. Finally, I would like to express my sincere gratitude to my family and friends for their support and encouragement.

Singapore
May 2024

Lei Wei

Contents

Chapter 1
Hollow-Core Antiresonant Fibers

Zhuo Wang, Mingjie Cui, and Changyuan Yu

Abstract Hollow-core fibers (HCFs) are special waveguides that can confine light waves in a low refractive index air region. They have much lower dispersion, nonlinearity, thermal sensitivity, and transmission delay than traditional solid-core fibers. At present, there are two types of HCFs, hollow-core antiresonant fibers (HC-ARFs) and hollow-core photonic band gap fibers (HC-PBGFs). Experiments have shown that HC-ARFs can achieve lower transmission loss and larger transmission bandwidth than HC-PBGFs. This chapter focuses on the development history, waveguide principle, loss mechanism, and advanced applications of HC-ARFs. At the same time, we compare the properties of HC-ARFs with those of HC-PBGFs. Section C introduces the origination and evolution of HC-ARFs, including different designs of fiber cross sections, as well as the advantages and disadvantages of different structures. Especially, the process of reducing losses of HC-ARFs is discussed in detail. The fundamental waveguide principle of HC-ARFs is introduced in Section D where the effects of antiresonant reflection and core-cladding mode phase mismatching are analyzed. Section E introduces three main factors causing loss in HC-ARFs: mode leakage, core-cladding mode coupling, and effects of cladding structures. The advanced applications of HC-ARFs including data transmission, supercontinuum generation, optical frequency comb distribution, and gas detection are introduced in Section F.

Z. Wang · M. Cui · C. Yu (✉)
Department of Electrical and Electronic Engineering, The Hong Kong Polytechnic University, Kowloon, Hong Kong, China
e-mail: changyuan.yu@polyu.edu.hk

Photonics Research Institute, The Hong Kong Polytechnic University, Kowloon, Hong Kong, China

Z. Wang
e-mail: z-eie.wang@connect.polyu.hk

M. Cui
e-mail: ming-jie.cui@connect.polyu.hk

Z. Wang
School of Information and Optoelectronic Science and Engineering, South China Normal University, Guangzhou 510006, China

L. Wei (ed.), *Advanced Optical and Optoelectronic Fibers*, Advances in Optics and Optoelectronics, https://doi.org/10.1007/978-981-97-6218-7_1

1 Introduction

Hollow-core fibers (HCFs) provide a unique platform for light guidance in an air core. Since most energy of fiber modes is confined in air, the interaction between light and the surrounding high-index materials is very weak. Therefore, optical modes supported in HCFs have the advantages of ultra-low Rayleigh scattering, minimal chromatic dispersion, weak nonlinearity, and small time delay. In the 1960s, it was suggested that hollow metallic and dielectric pipes could be used for optical transmission [1]. This pioneering work has shown that it is possible to use hollow waveguides to achieve long-distance light transmission. Notable experimental progress has been made over the past three decades, starting with the fabrication of the first hollow-core photonic bandgap fiber (HC-PBGF) [2]. The fiber enabled light transmission at the centimeter scale, attracting interest in the research of HCFs. A few years later, a class of hollow-core anti-resonant fibers (HC-ARFs) also drew intensive attention because of its wide transmission bandwidth. To date, numerous studies have focused on optimizing HC-PBGFs and HC-ARFs to achieve the lowest possible transmission loss. Although both HC-PBGFs and HC-ARFs are crucial in the field of HCFs, the content in this chapter will primarily delve into the advancements and developments in HC-ARFs.

The first experimental work using the antiresonant mechanism to realize optical guidance in HCFs was published in 2002 [3]. As shown in Fig. 1a, the cladding of the HCF is constructed by alternate elements from a hexagonal grid and a triangular grid, which is also called Kagome lattice. It can achieve a much wider transmission bandwidth in the visible and near-infrared bands compared with the first reported HC-PBGF [2]. Despite a loss peak at around 1390 nm, the transmission loss is lower than 3 dB/m in the wavelength range from 350 to 1700 nm. In the same year, a simple analytical theory for low-index core photonic crystal waveguides under the mechanism of antiresonant reflection was first proposed [4]. The model reveals that the guidance mechanism of antiresonant reflection is determined by the index contrast between the low-index and the high-index materials and the thickness of the high-index layer. At the same time, increasing the number of antiresonant layers can improve the confinement ability. However, this theoretical model was not linked to the experimental results of the hollow-core Kagome fiber (HC-KF) at that time. In 2007, HC-KFs were first used for the generation and guidance of a three-octave spectral comb, spanning wavelengths from 325 to 2300 nm [5]. The transmission loss was improved to ~0.5 dB/m. It was found that the high refractive index grids of cladding can be regarded as thin resonators. The antiresonant reflection mechanism can forbid the coupling between the core and cladding modes, which is key to decreasing the leakage loss of core modes.

So far, HC-KFs have gone through several generations of improvements. Firstly, increasing the pitch of the Kagome lattice was found can increase the transmission band and decrease the overlap between the core and interface modes. Figure 1b shows the cross section of a large-pitch (~12 μm) HC-KF [6]. This fiber has two transmission wavelength bands: one is in the visible wavelength range and has a bandwidth over

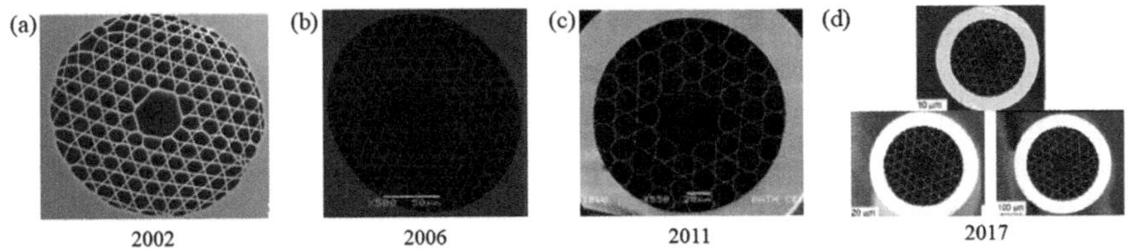

Fig. 1 The developments of HC-KFs: **a** The initial work on HC-KFs with loss < 3 dB/m in a wide wavelength band from 350 to 1700 nm (Adapted with permission from [3]. **b** The large-pitch HC-KF with an average transmission loss of around 0.75 dB/m in the near-infrared band (Adapted with permission from [6] © Optica Publishing Group). **c** The large-pitch hypocycloid-core HC-KF achieves low losses of ~ 180 dB/km over a transmission band larger than 200 THz (Adapted with permission from [7]). **d** Low-loss HC-KFs (Adapted with permission from [8] © Optica Publishing Group) with core diameters of 43 μm (top), 64 μm (bottom left), and 97 μm (bottom right)

250 μm and the other one is in the infrared band with a width of 700 nm. A flat transmission loss around 0.75 dB/m is in the near-infrared band and the minimum loss is 0.3 dB/m. In the visible band, the loss remains below 2 dB/m. Secondly, it was found that a hypocycloid core profile can help to improve the coupling inhibition between the core and cladding modes [7]. This discovery offers a new way to further reduce the loss of HC-KFs. The image of the first reported hypocycloid-core HC-KF is shown in Fig. 1c. The fiber has a lattice pitch of 20 μm, a core diameter of 50 μm, and a grid thickness of 280 nm. Taking advantage of the hypocycloid core profile, a low transmission loss (~180 dB/km) was realized in a transmission band larger than 200 THz. Thirdly, with the improvement of the fabrication technique, a series of low-loss HCKFs were reported in 2017 [8]. There are three hypocycloid-core HC-KFs with different core diameters and silica capillary thicknesses shown in Fig. 1d. The fiber shown on the top of Fig. 1d has a core diameter of 43 μm and a silica capillary thickness of 375 nm. Its transmission window is around 1000 nm with a 3 dB bandwidth of 150 nm. The minimum loss is 12.3 dB/km at the wavelength of 1010 nm. The structure shown on the bottom left has a larger core diameter of 64 μm and a thicker silica layer of 600 nm surrounding the core. Due to the antiresonant condition, the transmission window moves to around 1560 nm with a bandwidth of 198 nm. The minimum loss is 13.9 dB/km at the wavelength of 1566 nm. For the sample on the bottom right, the core diameter and silica capillary thickness are 97 μm and 975 nm, respectively. It has the lowest loss of 9.6 dB/km at the wavelength of 2462 nm. Although the HC-KFs have ultra-wide transmission bandwidth, their achievable minimum loss is hard to be further reduced. There will inevitably be thickness fluctuations at the lattice nodes. The scattering caused by nodes and the coupling between core and node modes are the main obstacles to reducing the loss.

The antiresonant reflection mechanism does not rely on multiple layers of air holes to reduce losses. The cladding structure of HC-ARFs can be further simplified to a single layer of negative curvature tubes. HCFs consisting of a single layer of dielectric cladding tubes were called tube lattice fibers in the original stage [9]. The structures of the multi-layer tube lattice and the single-layer tube lattice are shown in

Fig. 2a, b, respectively. To realize light confinement in the hollow core, the thickness of the tubes should meet the antiresonant condition. In addition, a larger core size is beneficial to decrease transmission loss because of the reduction in the overlap between the mode field and the dielectric material. However, a larger core size will lead to the guidance of higher-order modes, which will decrease the output beam quality. A way to increase the loss of the higher-order mode without affecting the loss of the fundamental mode is coupling the higher-order modes with cladding modes. It was found that, compared with the multi-layer tube lattice structure, the single-layer tube lattice structure is easier to couple the higher-order mode with the cladding mode while maintaining a low loss of the fundamental mode [9]. In 2011, the first experimental work on the single-layer tube lattice HC-ARF demonstrated the waveguiding ability in the spectral region larger than 3.5 μm [10]. Later, a HC-ARF with cone-type cladding that can realize a low loss of 34 dB/km at the wavelength of 3050 nm was reported [11]. An improved design and fabrication of nodeless HC-ARF was reported in 2013 [12]. This work showed that the fabricated fiber can guide a few modes with a measured loss of about 50 dB/km at the wavelength of 3390 nm. As shown in Fig. 2c, the cladding tubes of the nodeless HC-ARF are not connected. This design can eliminate the surface tension forces between the cladding tubes during the fiber drawing, which helps to keep the circular tube geometry even at a high drawing temperature. Besides, the nodeless structure eliminates the link between the cladding tubes, which reduces the generation of complex glass modes, and then eliminates the extra loss caused by Fano resonances [13, 14]. It should be noted that, because of the relatively thick tubes of the fabricated fiber, the early experiments on the HC-ARFs mainly demonstrated low transmission loss in the middle infrared wavelength range (>3000 nm). With the improvement of fabrication techniques, thinner tubes of HC-ARF can be fabricated with high quality. Later, the low-loss transmission band shifts to a shorter wavelength region. The experimental records of transmission loss in the nodeless single-layer-tube HC-ARF are 7.7 dB/km at the wavelength of 750 nm and in the range of 10–20 dB/km spanning from 600 to 1200 nm [15].

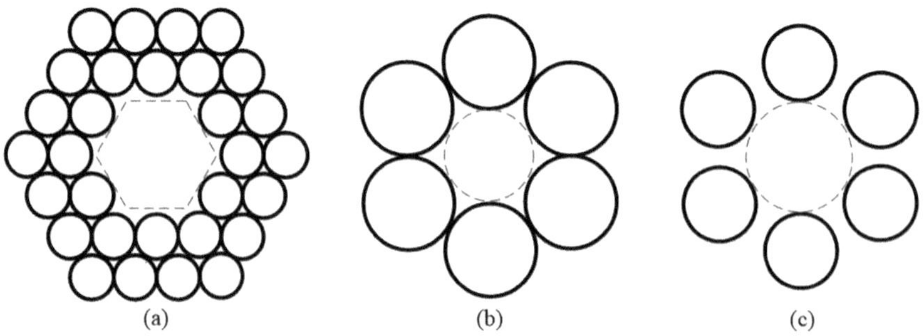

Fig. 2 Schematic diagrams of **a** multi-layer tube lattice, **b** single-layer tube lattice, and **c** nodeless single-layer tube lattice

To further decrease the loss of HC-ARFs, a nested-tube cladding structure was first proposed in 2014 [16]. A smaller tube is nested in the bigger tube without additional connection nodes. Therefore, there are two hypocycloid antiresonant layers to enhance the confinement of light. And the absence of additional nodes will remove extra scattering loss. The simulation work predicted that the nested-tube HC-ARF can achieve ultra-low confinement loss and scattering loss around 0.2 dB/km, simultaneously. The result first indicated that a HC-ARF is possible to realize a lower loss than a HC-PBGF. The first experimental results of nested-tube HC-ARF [17] were reported in 2016. Although the simulated loss of the designed fiber was ~1 dB/km, the measured transmission loss of the fabricated fiber was over 100 dB/km attributed to the geometrical deviations of the nested tubes.

Recently, breakthroughs have been made in the reduction of loss of HC-ARFs as shown in Fig. 3. In 2018, a new kind of conjoined-tube HC-ARF was proposed [18]. The cross-section image of the conjoined-tube HC-ARF is shown in Fig. 3a. A silica sheet is placed in the cladding tube to form an additional layer of antiresonant reflection. In the simulation, the conjoined-tube cladding can improve the confinement loss by 20 dB/km lower than the single-layer-tube cladding. The confinement loss can be further decreased below the surface scattering loss by adding two silica sheets in the cladding tube. Experiments showed that the conjoined-tube HC-ARF has a minimum loss of 2 dB/km at 1512 nm and a transmission band from 1302 to 1637 nm with losses lower than 16 dB/km. In the same year, a 6-nested-tube HC-ARF (Fig. 3b) was reported to realize losses lower than 1.5 dB/km with a bandwidth of 65 nm and the minimum loss is at 1450 nm with the value of 1.3 dB/km [19]. In 2019, a 6-nested-tube HC-ARF (Fig. 3c) was used for interband short-reach data transmission [20]. Taking advantage of a wide transmission band (1240–1940 nm) and low loss of 6.6 dB/km at 1550 nm, the fiber demonstrated a penalty-free 100 Gb/s PAM4 data transmission from O- to L-bands over 1 km. In 2020, the gap between HC-ARFs and standard all-silica single-mode fibers was further reduced. A 1.7-km long 6-nested-tube HC-ARF was fabricated with a high geometry quality (Fig. 3d) [21]. It has a low-loss transmission window from 1510 to 1600 nm with the lowest loss of 0.28 dB/km. In 2021, several 5-nested-tube HC-ARFs (Fig. 3e) with different fiber parameters were reported [22]. Ultra-low loss at wavelengths of 850, 1060, 1300, and 1625 nm was realized with the minimum loss of 0.22 dB/km at 1625 nm. Moreover, the loss was further decreased by using a double-nested HC-ARF in 2022 (Fig. 3f) [23]. The lowest loss was 0.174 dB/km in the C-band, which matches the loss of commercial solid-core fibers. In addition, the O-band loss was significantly reduced to 0.22 dB/km. In 2024, the recorded low loss of 0.08 ± 0.03 dB/km was demonstrated based on an optimized double-nested HC-ARF (Fig. 3g) [24], the attenuation is even lower than the lowest loss demonstrated in pure silica core single mode fibers.

The excellent performance mentioned above fully shows that HC-ARFs have a wide application prospect in the fields of optical communications, fiber sensing, and fiber laser. This chapter will introduce HC-ARFs in three parts, including the waveguide principles, the loss mechanisms, and the advanced applications.

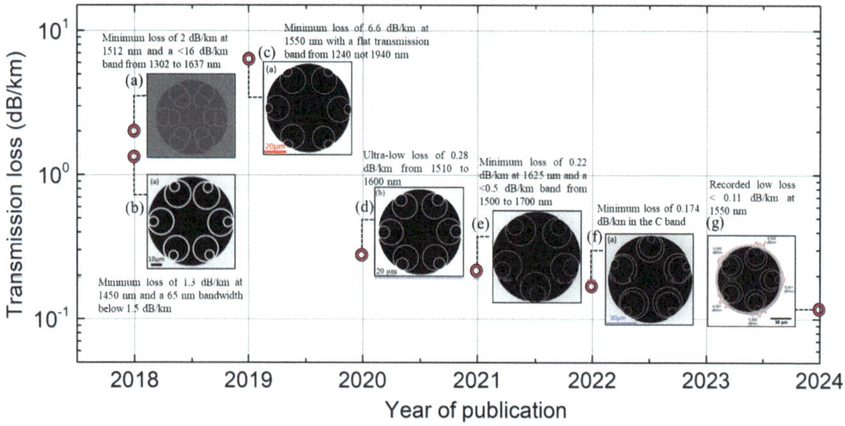

Fig. 3 The timeline of the recent progress of HC-ARFs. **a** The conjoined-tube HC-ARF has three silica layers with thicknesses corresponding to the second antiresonant window at the wavelength around 1450 nm and achieves the measured minimum loss of 2 dB/m at 1512 nm (Adapted with permission from [18]). **b** The nested-tube HC-ARF has losses lower than 1.5 dB/km with a bandwidth of 65 nm, and the minimum loss is at 1450 nm with a value of 1.3 dB/km (Adapted with permission from [19]). **c** The nested-tube HC-ARF with 700 nm bandwidth (1240–1940 nm) and a minimum loss of 6.6 dB/km at 1550 nm (Adapted with permission from [20] © Optica Publishing Group). **d** The nested-tube HC-ARF has a low loss of 0.28 dB/km from 1510 to 1600 nm (Adapted with permission from [21] © Optica Publishing Group). **e** The nested-tube ARF has a low loss of 0.22 dB/km at 1625 nm (Adapted with permission from [22] © Optica Publishing Group). **f** The double-nested HC-ARF has ultra-low losses of 0.174 dB/km and 0.22 dB/km in the C-band and O-band, respectively (Adapted with permission from [23] © Optica Publishing Group). **g** The optimized double-nested HC-ARF achieves the recorded low loss of 0.08 ± 0.03 dB/km (Adapted with permission from [24] © Optica Publishing Group)

2 Waveguide Principles

The light guiding mechanism in HC-ARFs is different from the total internal reflection in conventional solid-core fibers. To confine light in a lower refractive index region, the conditions of antiresonant reflection and inhibited mode coupling should be satisfied at the same time.

- Antiresonant Reflection

The high-index wall of cladding tubes can be regarded as a Fabry–Perot resonator. Therefore, when its thickness meets the condition of destructive interference, the leakage of the core mode can be effectively reduced. As shown by a simplified two-dimensional (2D) model in Fig. 4, the light enters the core-cladding interface with a glancing angle. In this case, the longitudinal component k_z, can be approximated by $n_0 k_0$. [4]. Thus, the angle of incidence is approximately equal to 90°. By using Snell's law, the transverse propagation constant in the silica layer can be expressed as $k_T = k_0\sqrt{n^2 - n_0^2}$. The thin wall of the silica tube can be considered as a cavity.

After double reflections, the two transmitted light beams at the outer boundary of the silica wall have a phase difference $\varphi_1 - \varphi_0 \approx 2tk_T$. For $\varphi_1 - \varphi_0 = 2m\pi$, where m is a positive integer, the two light beams have a constructive interference. While, in the condition $\varphi_1 - \varphi_0 = (2m-1)\pi$, the two light beams have a destructive interference. Thus, it can be concluded that:

$$t = \frac{m\lambda}{2\sqrt{n^2 - n_0^2}}, \text{ (Resonant)}, \tag{1}$$

$$t = \frac{(2m-1)\lambda}{4\sqrt{n^2 - n_0^2}}, \text{ (Antiresonant)}, \tag{2}$$

where t is the tube wall thickness, n and n_0 are the refractive indexes of silica and air, respectively. In the antiresonant condition, light experiences destructive interference in the Fabry–Perot-like cavity, which leads to low leakage from the air core to the silica tubes. Thus, optical modes can be well confined in the air core.

For a fixed thickness of tubes, there are several transmission windows in a wide wavelength band. The bandwidth of a transmission window decreases with the increase of the order of antiresonant conditions. A coefficient $M = 2t\sqrt{n^2 - n_0^2}\Big/\lambda$

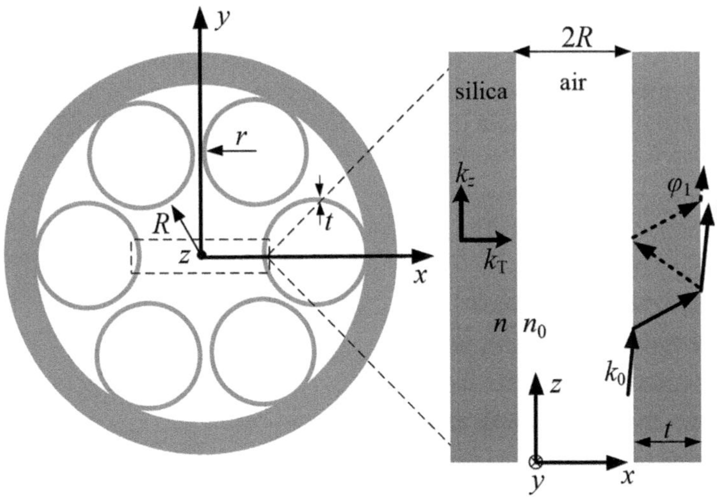

Fig. 4 Schematic diagram of antiresonant mechanism. Core radius, cladding tube radius, and tube wall thickness are represented by R, r, and t, respectively. The right panel shows the section of the hollow core along the z direction. The magnitude of the wave vector in the air region is denoted by k_0. The propagation constant in the silica region is decomposed into transverse (k_T) and longitudinal (k_z) components. The phases of two successive beams of transmitted light at the outer boundary of the silica layer are denoted as φ_0 and φ_1

is defined to describe if the operation wavelength is in the condition of antiresonance or resonance for fixed structural parameters. Figure 5 shows M versus wavelength under different tube wall thicknesses. The resonant bands exist where M is a positive integer. There is a low-loss anti-resonant transmission window below the first-order resonant band or between two adjacent resonant bands. From Fig. 5a we can find that the thinner the tube wall, the wider the anti-resonant transmission window can be achieved in the near-infrared band. Figure 5b illustrates the numerically calculated confinement loss of a 6-tube HC-ARF. The core radius and tube inner radius are 15 μm and 12.75 μm, respectively. When t is 0.42 μm, the confinement loss remains below 1 dB/m in a wide wavelength band from 0.95 to 1.8 μm. However, when t increases to 1.159 μm, the low-loss window is split into two parts from 0.9 to 1.075 μm and from 1.3 to 1.75 μm, respectively. The low-loss window obtained by the numerical method is consistent with the results predicted by the analytical model.

- Core-Cladding Mode Phase Mismatching

In a HC-ARF, the cladding tubes can be regarded as hollow waveguides with high loss. When a core mode has a n_{eff} approaching that of modes supported in the cladding tubes, the coupling between the core and tube modes is enhanced by phase matching. It can cause a high loss of the core mode. The mode patterns in a 6-tube HC-ARF are shown in Fig. 6. For the HE_{11} mode, its effective index n_{eff} is different from that of modes in the cladding tubes. So, the core-cladding mode coupling is inhibited because of phase mismatching. Consequently, HE_{11} mode can be well confined in the core region. However, for the HE_{21} mode, a severe coupling happens caused by phase matching as shown by the mode profile in Fig. 6b. Although the antiresonant condition is met, the mode coupling causes most of the model field exists in the cladding tubes, which results in high modal loss. Therefore, the ratio r/R should be designed carefully to avoid coupling the desired core mode into the cladding [25].

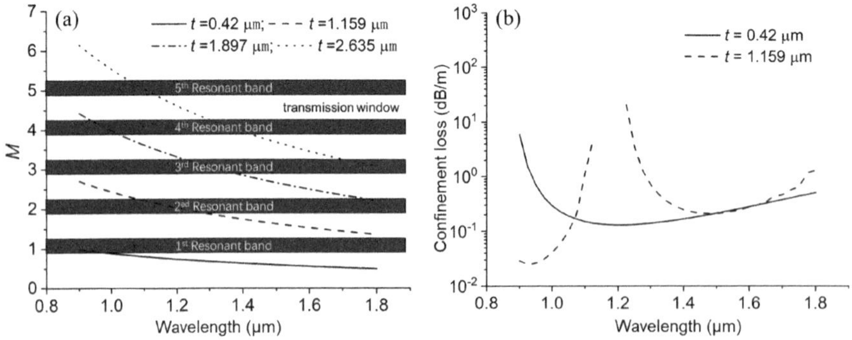

Fig. 5 **a** The coefficient of resonance M versus wavelength with different tube wall thicknesses. **b** The confinement loss of the fundamental mode in a 6-tube HC-ARF with different t. The core radius and tube inner radius are 15 μm and 12.75 μm, respectively

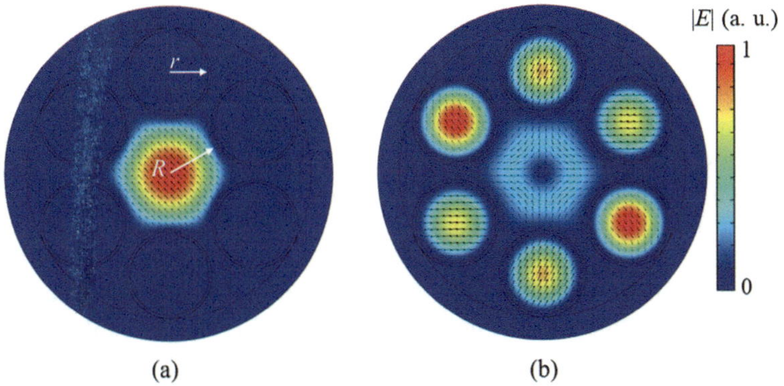

Fig. 6 The model fields of **a** HE_{11} and **b** HE_{21} modes in a 6-tube HC-ARF when $r/R = 0.7$ at the wavelength of 1.55 μm. The black arrows indicate the direction of polarization

On the other hand, core-cladding mode coupling can be used to eliminate unwanted higher-order modes to achieve two-mode transmission with high mode purity [26].

In this section, we introduced two fundamental waveguide principles of a HC-ARF. First, the antiresonant reflection is to prevent the energy leakage from the core to the cladding through the principle of destructive interference. Second, the phase mismatching of core and cladding modes inhibits the core-cladding mode coupling, which is also vital for the transmission of core modes. However, when the two conditions are met, it is still difficult for a single-layer tube lattice HC-ARF to achieve ultra-low transmission loss. Therefore, the loss mechanisms and more complex structures of HC-ARFs will be introduced in the next section.

3 Loss Mechanisms

Different from the conventional optical fibers, the loss mechanism in HCFs is more complicated. For the complex structure so far, the loss contains leakage loss, coupling loss, scattering loss, material-related absorption loss, and some extrinsic losses caused by external perturbations. Here, for the fundamentals of HC-ARFs, we give a brief introduction to the concept of leakage loss, mode coupling, and cladding design.

- Leakage loss

From the perspective of ray optics, light propagating in an optical fiber obeys the rule of total internal reflection. This kind of reflection will not suffer from power loss, which means, theoretically, rays can travel along the fiber forever. However, when the propagating modes are below their cutoff or the fiber is subjected to some perturbations such as bending, cases become very different. In those cases, the reflection of rays is not perfect anymore but suffers from power loss. Some rays will leak out when each reflection occurs on the interface of the core and the cladding. Those rays leaked out from the fiber core can be understood as leakage loss. When it comes to waveguide modes, one can see that the bound modes are characterized by a real propagation constant. However, for modes below their cutoff or in perturbed waveguides they can possess a propagation constant with an imaginary part. The propagation constant is a complex number in these cases and the value of its imaginary indicates the amount of power loss when the modes are propagating along the waveguide. Here the modes with complex propagation constant are called leaky modes [27]. As a representation of leakage loss, the imaginary part of the propagation constant is mostly presented in the Decibel unit in the numerical analysis and the conversion formula is as follows:

$$\alpha_{Loss} = \frac{20}{\ln 10}\mathrm{Im}(\beta). \tag{3}$$

For bound modes, we can find $\alpha = 0$ in high-index nonabsorbing waveguides. In hollow-core waveguides, the air core region is surrounded by high-index material, such as silica, and structures like that cannot support any bound modes because the modes found in the core region possess an effective index always lower than the index of outer silica. So, modes propagating in the core region are known as leaky modes.

In Fig. 7, the wavelength-dependent leakage loss of two simple hollow-core waveguides is shown. For the hollow core formed by an infinite large silica one can see that the modes suffer from a leakage loss that cannot be neglectable. Higher-order modes suffer from a higher leakage loss as shown in Fig. 7 by the dash lines. An intuitive explanation is that the reflectivity gets lower because the incident angles corresponding to the higher-order modes are more oblique when the rays enter the interface of the core and cladding. Moreover, it can be seen from the curves that the leakage loss increases with wavelength. This is because the modal field gets larger with the increasing wavelength. As a result, the amount of leakage loss increases. For the hollow core formed by a circular layer, lower leakage loss can be obtained owing to the one-layer structure regardless of the resonant peaks [29].

- Loss caused by mode coupling

Mode couplings are common phenomena in a waveguide system. Given two identical parallel circular waveguides, optical power will be transferred from one waveguide to another periodically when the two waveguides are close to each other. This is an obvious characteristic of a complete mode coupling. For the tube structure shown in Fig. 7, the resonant peaks that cause high leakage loss can be considered a result of mode coupling between the air-core mode and the silica mode. For HCFs with complex structures such as HC-ARFs, their cladding structure can be regarded as a combination of multiple waveguides. These waveguides support numerous modes with certain propagation constants, which increases the probability of the core mode coupled to the cladding modes [15]. Thus, huge losses can be caused by the mode couplings between core and cladding modes. The additional loss due to the mode

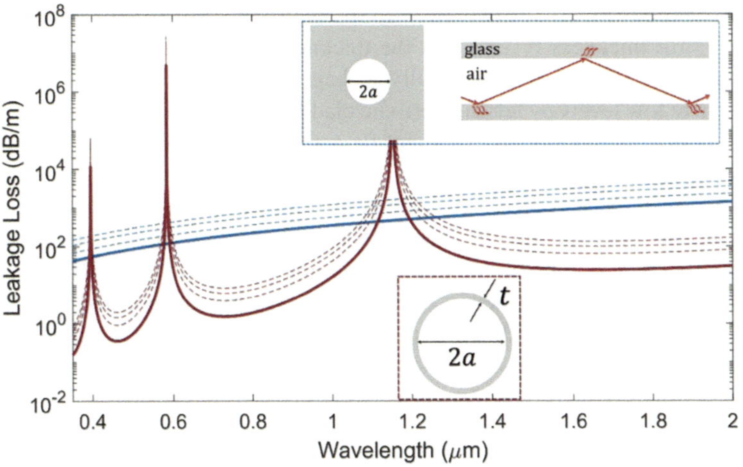

Fig. 7 Leakage loss of two simple hollow-core waveguide structures. Blue curves and red curves represent the losses of the hollow dielectric waveguide and dielectric tube waveguide, respectively (Adapted with permission from [28] © Optica Publishing Group)

couplings between core mode i and cladding mode j can be approximated by the following expression:

$$\alpha_{i,j} \propto \frac{\gamma |k_{ij}|}{(\Delta\beta)^2}, \tag{4}$$

where γ_j indicates the loss of the cladding mode, κ_{ij} the coupling coefficients between the two modes and $\Delta\beta$ is the difference between propagation constants. One can see that if a low-loss transmission of core modes is desired, the couplings must be inhibited to prevent the power of core modes from leaking into the cladding [30]. Specifically, by coupled mode formalism, one can learn that mode coupling depends on two key parameters, the difference between propagation constants $\Delta\beta$ and the coupling coefficient κ_{ij}. For the reduction of mode coupling, it is better to have the absolute value of $\Delta\beta$ sufficiently large and make the κ_{ij} as small as possible. Further, even if the phase match condition is satisfied ($\Delta\beta = 0$), the effective inhibited coupling can be also achieved by assuring the coupling coefficient κ_{ij} equals zero.

- Leakage loss affected by cladding design

The cladding structure plays an important role in determining the level of leakage loss. In the simple HCFs case, it is already shown that the leakage loss can be reduced significantly by increasing the diameter of the core region. Also, by adding more cladding layers the loss can reach extremely low levels. However, some limitations prevent us from getting lower losses. In the previous study of HC-PBGFs [28], it was found that even for structures with the same number of layers, their loss spectra are very different in magnitude. The calculated loss spectrum by the finite element method (FEM) for two HC-PBGFs with the same structure but different cladding strut thickness reveals that the thickness of the cladding strut also has a great impact on leakage loss. Generally, leakage loss of HC-PBGFs can be reduced to arbitrarily low levels by adding periodic cladding structures. However, it is found that the effect of cladding struts cannot be neglected. Thicker cladding struts may support an increasing number of modes with a defined effective index, which will reduce the difference between core and cladding modes in propagation constant improving the level of coupling between core and cladding modes. Therefore, high leakage loss could be introduced in this case. For single-ring HC-ARFs, a negative curvature hollow core is formed by the surrounding tubes. Such a cladding structure can contribute to reducing the resonant peaks, providing a wideband transmission window. Moreover, without connected nodes between the tubes, HC-ARFs do not seem to be affected by the cladding strut shown in HC-PBGFs. However, the leakage loss of HC-ARFs can still be easily affected by the nodes between tubes and the outer jacket and by the thickness of layers.

In Fig. 8, we can see the leakage loss is reduced greatly in the fiber with nested tubes. Note that the two fibers have the same core diameter, tube gaps, and wall thickness. In addition, by careful design of the ratio of core diameter to tube diameter one can reduce the propagation constant difference between higher-order core modes and cladding modes and increase their coupling, which means that broadband single-mode operation can be realized in HCFs. Nevertheless, the cladding design of HC-ARFs can be further improved by optimizing the number of tubes.

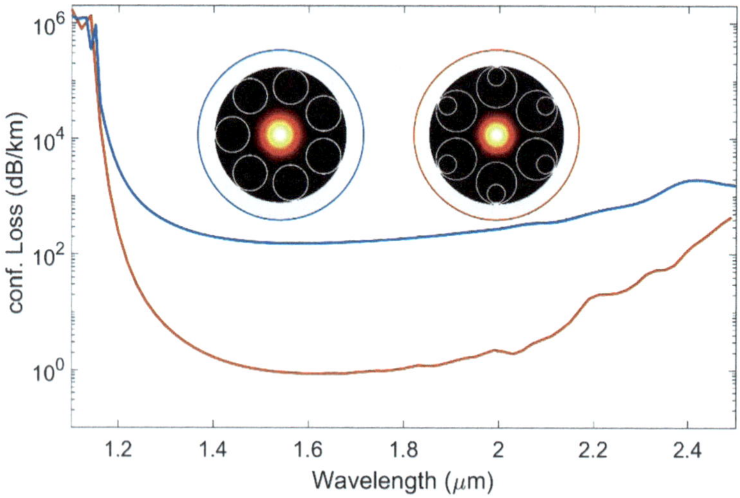

Fig. 8 Leakage loss (confinement loss) of two different HC-ARFs. Blue curves and red curves represent the losses of the single-ring antiresonant hollow-core fiber and the nested antiresonant nodeless hollow-core fiber, respectively (Adapted with permission from [28] © Optica Publishing Group)

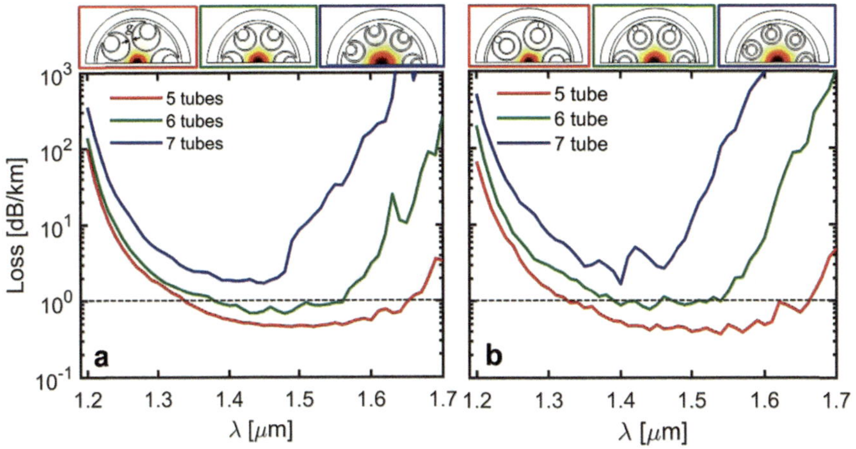

Fig. 9 Calculated loss spectra of regular **a** and modified **b** nested HC-ARFs with five, six, and seven tubes (Adapted with permission from [31] © Optica Publishing Group)

As shown in Fig. 9, the HC-ARFs with five cladding tubes show lower leakage loss and wider transmission window than the other tube numbers. This result is not only for the nested HC-ARFs but also for its modified type. It has been pointed out that a five-tube structure provides a larger distance between the core surround and the outer tube compared with the structures containing six and seven tubes, and thus, it supports a lower loss and broader transmission window.

In addition to the optimization of the number of cladding tubes, we need to focus on the shape of the tubes. As illustrated in Fig. 10a, HC-ARFs with elliptical tubes show lower confinement loss compared to HC-ARFs with circular tubes in both the single-ring and nested cases in the wavelength range from 0.9 to 1.7 μm. Moreover, the lowest loss can be obtained by an HC-ARF nesting elliptical tube inside a circular tube compared to the other four HC-ARFs. It is not difficult to find that elliptical tubes perform better than circular tubes in obtaining lower losses in these cases. Although the shape of cladding elements affects the leakage loss observably, the mechanism beneath it is still not clear. Recently, a concept of azimuthal confinement introduced by Murphy and Bird provides a new perspective for further understanding the loss mechanism in hollow-core antiresonant fibers [33]. As shown in Fig. 10b, by shaping the wave field in the azimuthal coordinate the wave functions within the tubes are strongly confined. For elliptical tubes, the azimuthal confinement operates over larger distances, which offers the strongest confinement and leads to a lower loss. Nevertheless, the theoretical model developed in the paper also implies that negative curvature of the core surround may play no intrinsic role in reducing the leakage loss in the fibers with nodeless cladding structure. The discovery of azimuthal confinement reveals that though negative curvature of the core surround is always correlated with low loss, azimuthal confinement plays a crucial role in determining the confinement loss. The emergence of the concept of azimuthal confinement will

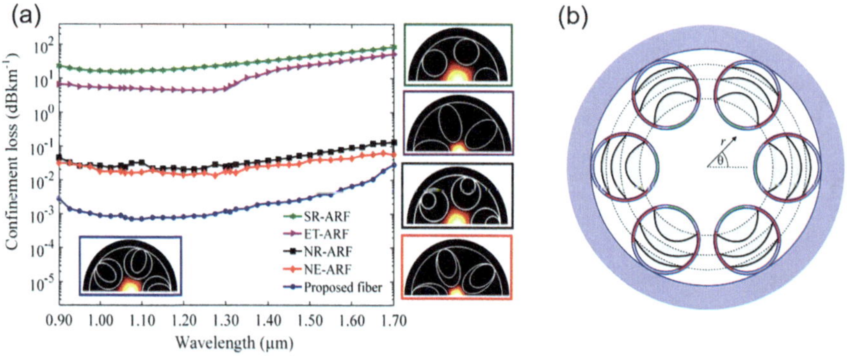

Fig. 10 **a** Calculated loss spectra of a single ring antiresonant fiber (SR-ARF), an elliptical tube antiresonant fiber (ET-ARF), a nested ring antiresonant fiber (NR-ARF), a nested elliptical antiresonant fiber (NE-ARF) and the proposed fiber in literature (Adapted with permission from [32]). **b** Schematic illustration of azimuthal confinement in a single ring fiber (Adapted with permission from L. R. Murphy and D. Bird 2023 © Optica Publishing Group)

contribute to the interpretation of loss mechanisms in many HC-ARFs and the design of complex hollow-core waveguides.

4 Applications

HC-ARFs have the characteristics of low loss, weak nonlinear effects, small chromatic dispersion, and low transmission latency. They can be used for optical communications and fiber lasers. In addition, by filling the hollow core with target gas, one can enhance the light-molecule interaction, achieving highly sensitive gas detection. This section will mainly introduce HC-ARFs from the perspective of advanced applications.

- Data Transmission

Transmission latency and nonlinear effect accumulation are two difficult issues faced by conventional fiber-optic communication networks. Ideally, the refractive index of air is 1 and there are almost no nonlinear effects in vacuum. Thus, HCFs have the potential to fundamentally address both challenges mentioned above. Different kinds of HCFs have been tried to conduct low-latency data transmission. In 2013, a large-core HC-PBGF was first reported to achieve low-latency data transmission [34]. This work is the first experimental demonstration of fiber-based wavelength-division-multiplexing (WDM) data transmission at close to (99.7%) the speed of light in the vacuum. After that, a 310 m long HC-PBGF with a 37-cell structure was reported enabling WDM and mode-division-multiplexing (MDM) transmission with a high capacity of 73.7 Tb/s [35]. The HC-PBGF can support LP_{01} and LP_{11} modes with low loss and weak modal crosstalk, and the minimum loss of 3.3 dB/km at 1550 nm and an 85 nm wide 3-dB bandwidth were demonstrated. In 2015, a loop experiment over HC-PBGF with a length of 74.8 km using 40G dual-polarization quadrature phase shift keying (DP-QPSK) modulation format was reported. The results indicate that the HCFs are good quasi-linear media that can realize high-power signal transmission [36]. Although HC-PBGFs can achieve low-loss optical signal transmission, its low-loss transmission bandwidth is difficult to be further broadened. Different from HC-PBGFs, HC-ARFs have ultra-wide low-loss transmission windows. Hence, they are more competitive for high-capacity data transmission. In 2017, data transmission at wavelengths nearly one octave apart based on a single-layer tube lattice HC-ARF was reported for the first time [37]. In the wavelength range from 785 to 1705 nm, the measured transmission loss was < 100 dB/km. This HC-ARF was shown to support penalty-free data transmission through 100 m at the wavelengths of 1065 nm, 1565 nm, and 1963 nm. In addition, the ultra-low nonlinearity gives HC-ARFs a great advantage in high-power data transmission. It was demonstrated that a single-layer tube lattice HC-ARF can support single-mode transmission with a high launch power of 32.5 dBm, resulting in an output power of 28.5 dBm after 100-m transmission [38]. In the case of similar output power (28.5 dBm), the HC-ARF has a 2 dB signal-to-noise ratio (SNR) advantage over a commercial single-mode fiber (SMF-28). In another recent work, systematic analyses were made for a conjoined-tube HC-ARF, a single-layer tube lattice HC-ARF, and a HC-PBGF [39]. The conjoined-tube HC-ARF has a lower transmission loss (3.7 dB/km at 1550 nm) than the other two fibers. Under a 10 Gbit/s on–off keying (OOK) transmission test, the 200-m conjoined-tube HC-ARF and 109-m single-layer tube lattice HC-ARF have similar performance. Lower power penalties can be obtained by the two HC-ARFs than the HC-PBGF. The above results show that HC-ARFs with low loss, weak nonlinear, and small time delay are expected to achieve high-power data transmission in realistic environments.

- Supercontinuum Generation

Hollow-core waveguides filled with noble gas are ideal media for ultrafast supercontinuum generation. HC-ARFs can support single-mode transmission with a much larger mode diameter than commercial single-mode fibers, which is beneficial for achieving a higher total power of supercontinuum. In 2017, a HC-ARF with a core diameter of 50 μm was reported for ultrafast supercontinuum generation [40]. On the cross section of the HC-ARF, the antiresonant silica wall forms the shape of a six-point star, so the central hollow core is hexagonal. When femtosecond pulses launched into the krypton-gas-filled HC-ARF, the output spectrum was effectively broadened into a wideband from 200 to 1700 nm at an output energy of ~23 μJ within a single mode. It was explained that the structural resonances in the HC-ARF offer an additional degree of freedom to manage dispersion. The group velocity dispersion landscape at the strand resonance condition changes from anomalous to normal or from normal to anomalous with several zero-dispersion points. Numerical simulations suggested that the strongly varying group velocity dispersion outperforms the dispersion of gas. It resulted in supercontinuum generation under the mechanisms of coherent non-adiabatic soliton dynamics and multiple Cherenkov radiation generation events.

- Optical Frequency Comb Distribution

The biggest disadvantage of solid-core fibers used in optical frequency comb (OFC) distribution is the thermal sensitivity of chromatic dispersion, which can significantly reduce the accuracy of the transmitted OFC in its extended bandwidth. As the thermal-induced chromatic dispersions vary with wavelength, the optical path compensation technique cannot provide a high level of stability across the entire optical spectrum of the OFC. Low-loss HC-ARFs guide light in the central hole instead of silica, providing ultra-low thermal sensitivity, chromatic dispersion, and nonlinearity, which is ideal for the stable transmission of OFC through several kilometres-long distances. In 2022, a nested-tube HC-ARF was used for stable OFC transmission over 7.7 km [41]. The measured OFC frequency stability was 4.5×10^{-17} at 2000s over a 60–120 nm OFC bandwidth. In addition, the transfer frequency stability of the optical mode and mode spacing was 1.8×10^{-19} and 1.5×10^{-17} at a few thousand seconds, respectively. These good performances show that HC-ARFs are excellent media for OFC distribution both within the laboratory and indeed to remote external users.

- Gas Detection

HC-ARFs can be used as promising gas cells for high-precision gas detection systems. In 2020, highly precise gas detection under the mechanism of mode-phase-difference (MPD) based on a dual-mode HC-ARF was reported [42]. The HC-ARF can guide LP_{01} and LP_{11} modes with low loss. The pump-caused thermal-induced refractive index varies periodically along the gas-filled fiber because of the two-mode interaction. As the probe light also consists of LP_{01} and LP_{11} modes, the photothermal phase modulations on the two modes are different. However, the effect of external disturbance on the phase of the two modes is almost the same, resulting in the MPD being insensitive to environmental perturbation. The experiment demonstrated acetylene detection with ultra-high precision of parts-per-trillion and the instability was reduced to less than 1% over 3 hours.

5 Conclusion

In this chapter, the HC-ARFs are introduced from four aspects including the history of development, waveguide principle, loss mechanism, and advanced applications. The basic idea of the antiresonant waveguide is to treat the thin silica wall as a Fabry–Perot resonator. Light propagating outward through the silica wall is prevented by destructive interference under the condition of grazing incidence from air to silica. This mechanism makes it possible to achieve a large low-loss transmission bandwidth better than HC-PBGFs. In addition, the transmission loss of HC-ARF can be further reduced by the design of nodeless and nested cladding structures. The phenomenon of core-cladding mode coupling is discussed. By carefully designing the ratio of the hollow core to the cladding tubes, the core-cladding mode coupling mechanism can be used to realize mode selection, which is very important for the design of few-mode HC-ARFs with high mode purity. Currently, the advanced manufacturing technology can achieve stable fabrication of kilometer-length low-loss HC-ARFs. The latest reported double-nested HC-ARF has achieved a low-loss record of 0.08 $\pm$ 0.03 dB/km with a length of 2.5 km. In addition to ultra-low loss, HC-ARFs have characteristics of weak dispersion, low nonlinearity, weak photothermal effect, and low transmission latency. Therefore, HC-ARFs can be used for important cutting-edge applications in the fields of data transmission, supercontinuum generation, optical frequency comb distribution, and gas detection.

References

1. Marcatili EAJ, Schmeltzer RA. Bell Syst Tech J. 1964;43:1783–809.
2. Cregan RF, Mangan BJ, Knight JC, Birks TA, St. J. Russell P, Roberts PJ, Allan DC. Science. 1999;285:1537–1539.
3. Benabid F, Knight JC, Antonopoulos G, St. J. Russell P. Science. 2022;298:399–402.
4. Litchinitser NM, Abeeluck AK, Headley C, Eggleton BJ. Opt Lett. 2002;27:1592–4.
5. Couny F, Benabid F, Roberts PJ, Light PS, Raymer MG. Science. 2007;318:1118–21.
6. Couny F, Benabid F, Light PS. Opt Lett. 2006;31:3574–6.
7. Wang YY, Wheeler NV, Couny F, Roberts PJ, Benabid F. Opt Lett. 2011;36:669–71.
8. Wheeler NV, Bradley TD, Hayes JR, Gouveia MA, Liang S, Chen Y, Sandoghchi SR, Abokhamis Mousavi SM, Poletti F, Petrovich MN, Richardson DJ. Opt Lett. 2017;42:2571–2574
9. Vincetti L, Setti V. Opt Express. 2010;18:23133–46.
10. Pryamikov AD, Biriukov AS, Kosolapov AF, Plotnichenko VG, Semjonov SL, Dianov EM. Opt Express. 2011;19:1441–8.
11. Yu F, Wadsworth WJ, Knight JC. Opt Express. 2012;20:11153–8.
12. Kolyadin AN, Kosolapov AF, Pryamikov AD, Biriukov AS, Plotnichenko VG, Dianov EM. Opt Express. 2013;21:9514–9.
13. Vincetti L, Setti V. Opt Express. 2012;20:14350–61.
14. Gao S, Wang Y, Liu X, Ding W, Wang P. Opt Express. 2016;24:14801–11.
15. Debord B, Amsanpally A, Chafer M, Baz A, Maurel M, Blondy JM, Hugonnot E, Scol F, Vincetti L, Gérôme F, Benabid F. Optica. 2017;4:209–17.
16. Poletti F. Opt Express. 2014;22:23807–28.
17. Kosolapov AF, Alagashev GK, Kolyadin AN, Pryamikov AD, Biryukov AS, Bufetov IA, Dianov EM. Quantum Electron. 2016;46:267–70.
18. Gao S, Wang Y, Ding W, Jiang D, Gu S, Zhang X, Wang P. Nat Commun. 2018;9:1–6.
19. Bradley TD, Hayes JR, Chen Y, Jasion GT, Sandoghchi SR, Slavik R, Fokoua EN, Bawn S, Sakr H, Davidson IA, Taranta A, Thomas JP, Petrovich MN, Richardson DJ, Poletti F. European conference on optical communication (ECOC), 2018;1–3.
20. Sakr H, Bradley TD, Hong Y, Jasion GT, Hayes JR, Kim H, Davidson IA, Numkam Fokoua E, Chen Y, Bottrill KRH, Taengnoi N, Petropoulos P, Richardson DJ, Poletti F. Optical fiber communications conference and exhibition (OFC), Th4A.1 (2019).
21. Jasion GT, Bradley TD, Harrington K, Sakr H, Chen Y, Fokoua EN, Davidson IA, Taranta A, Hayes JR, Richardson DJ, Poletti F. Optical fiber communication conference (OFC), Th4B.4. (2020).
22. Sakr H, Bradley TD, Jasion GT, Fokoua EN, Sandoghchi SR, Davidson IA, Taranta A, Guerra G, Shere W, Chen Y, Hayes JR, Richardson DJ, Poletti F. Optical fiber communications conference and exhibition (OFC), F3A.4 (2021).

23. Jasion GT, Sakr H, Hayes JR, Sandoghchi SR, Hooper L, Fokoua EN, Saljoghei A, Mulvad HC, Alonso M, Taranta A, Bradley TD, Davidson IA, Chen Y, Richardson DJ, Poletti F. Optical fiber communications conference and exhibition (OFC), Th4C.7 (2022).
24. Chen Y, Petrovich MN, Fokoua EN, Adamu AI, Hassan MRA, Sakr H, Slavík R, Gorajoobi SB, Alonso M, Ando RF, Papadimopoulos A, Varghese T, Wu D, Ando MF, Wisniowski K, Sandoghchi SR, Jasion GT, Richardson DJ, Poletti F. Optical fiber communications conference and exhibition (OFC), Th4A.8 (2024).
25. Uebel P, Günendi MC, Frosz MH, Ahmed G, Edavalath NN, Ménard J, St.J. Russell P. Opt Lett. 2016;41:1961–1964.
26. Wang Z, Tu J, Liu Z, Yu C, Lu C. J Light Technol. 2020;38:864–74.
27. Snyder AW, Love JD, Optical waveguide theory (Chapman and Hall, 1983).
28. Fokoua EN, Mousavi SA, Jasion GT, Richardson DJ, Poletti F. Adv Opt Photon. 2023;15:1–85.
29. Bird D. Opt Express. 2017;25:23215–37.
30. Wei C, Weiblen RJ, Menyuk CR, Hu J. Adv Opt Photon. 2017;9:504–61.
31. Habib MdS, Antonio-Lopez JE, Markos C, Schülzgen A, Amezcua-Correa R. Opt Express. 2019;27:3824–36.
32. Shaha KSR, Khaleque A, Hosen MS. J Light Technol. 2021;39(20):6585–91.
33. Murphy LR, Bird D. Optica. 2023;10:854–70.
34. Poletti F, Wheeler NV, Petrovich MN, Baddela N, Numkam Fokoua E, Hayes JR, Gray DR, Li Z, Slavík R, Richardson DJ. Nat Photon. 2013;**7**:279–284.
35. Sleiffer VAJM, Jung Y, Baddela NK, Surof J, Kuschnerov M, Veljanovski V, Hayes JR, Wheeler NV, Fokoua ERN, Wooler JP, Gray DR, Wong NH, Parmigiani FR, Alam S, Petrovich MN, Poletti F, Richardson DJ, de Waardt H. J Light Technol. 2014;32:854–63.
36. Kuschnerov M, Sleiffer VAJM, Chen Y, de Man E, Chen Y, Liu Z, Sandoghchi SR, Jasion GT, Bradley T, Fokoua EN, Hayes JR, Wheeler NV, Gray DR, Slavik R, Jung Y, Wong NL, Mangan BJ, Poletti F, Petrovich MN, Richardson DJ. European conference on optical communication (ECOC). 2015;1–3.
37. Hayes JR, Sandoghchi SR, Bradley TD, Liu Z, Slavík R, Gouveia MA, Wheeler NV, Jasion G, Chen Y, Fokoua EN, Petrovich MN, Richardson DJ, Poletti F. J Light Technol. 2017;35:437–42.
38. Liu Z, Karanov B, Galdino L, Hayes JR, Lavery D, Clark K, Shi K, Elson DJ, Thomsen BC, Petrovich MN, Richardson DJ, Poletii F, Slavík R, Bayvel P. J Light Technol. 2019;37:909–16.
39. Wang X, Ge D, Ding W, Wang Y, Gao S, Zhang X, Sun Y, Li J, Chen Z, Wang P. Opt Lett. 2019;44:2145–8.
40. Sollapur R, Kartashov D, Zürch M, Hoffmann A, Grigorova T, Sauer G, Hartung A, Schwuchow A, Bierlich J, Kobelke J, Chemnitz M, Schmidt MA, Spielmann C. Light Sci Appl. 2017;6: e17124.
41. Feng Z, Marra G, Zhang X, Fokoua ERN, Sakr H, Hayes JR, Poletti F, Richardson DJ, Slavík R. Laser Photonics Rev. 2022;16:2200167.
42. Zhao P, Zhao Y, Bao H, Ho HL, Jin W, Fan S, Gao S, Wang Y, Wang P. Nat Commun. 2020;11:847.

Chapter 2
Specialty Fiber Technology for Brillouin Optical Time Domain Analysis

Dora Juan Juan Hu, Hui Dong, and Hailiang Zhang

Abstract Brillouin optical time domain analysis (BOTDA) is a powerful technique for measuring the temperature and strain distribution along an optical fiber using Brillouin scattering. It has offered many opportunities in various industries including structural health monitoring for civil infrastructure, oil and gas industry, geophysical exploration, environmental monitoring etc. As the technology of specialty fibers is continuously evolving, it is bringing new and enhanced opportunities for BOTDA applications by enabling new sensing capabilities and addressing specific challenges. For example, specialty fibers with special materials and/or special structures have been applied in BOTDA system to achieve enhanced sensor performance, including large scale strain sensing, temperature-insensitive strain sensing, high temperature strain sensing. These performance enhancements lead to many favourable applications, such as curvature sensing and shape reconstruction, discriminative sensing of temperature and strain. Multiplexed sensing systems have been developed to realize multipurpose sensing capabilities, such as hybrid Brillouin and Rayleigh sensing system, hybrid Brillouin and Raman sensing system. This book chapter introduces the fundamentals of utilization of unique properties of specialty fibers in BOTDA system and highlights specific applications. This book chapter could serve as a useful resource for researchers, engineers, and scientists working in the field of optical sensing and fiber optics.

D. J. J. Hu (✉) · H. Dong · H. Zhang
Agency for Science, Technology and Research (A*STAR), Institute for Infocomm Research (I2R), 1 Fusionopolis Way, #21-01, Connexis South Tower, Singapore 138632, Republic of Singapore
e-mail: jjhu@i2r.a-star.edu.sg

H. Dong
e-mail: hdong@i2r.a-star.edu.sg

H. Zhang
e-mail: zhang_hailiang@i2r.a-star.edu.sg

L. Wei (ed.), *Advanced Optical and Optoelectronic Fibers*, Advances in Optics and Optoelectronics, https://doi.org/10.1007/978-981-97-6218-7_2

1 Introduction

Brillouin scattering, named after the French physicist Léon Brillouin, was theoretically predicted in 1922 [9]. The stimulated Brillouin scattering (SBS) is a nonlinear process dependent on input power. The backscattered Stokes light interferes with the input pump light and generates an acoustic wave through the effect of electrostriction [36]. The Brillouin frequency shift (BFS) is the frequency difference between the Stokes wave and the pump at which the amplification of the Stokes wave is the maximum. BFS depends on the mechanical and thermal conditions of the fiber, i.e., increasing with both strain and temperature, forming the sensing principle of the Brillouin optical time domain analysis (BOTDA) [36]. The SBS process in which the acoustic wave generated from the light beam was first observed in 1964 [11], and was later observed in optical fiber and reported by Ippen et al. in 1972 [21]. The SBS process was harnessed to develop Brillouin fiber sensors to measure physical parameters [36].

Since the introduction of BOTDA system in 1990 [65], it has undergone rapid development primarily owing to its ability to provide continuous and highly accurate distributed measurement of temperature and/or strain in optical fibers [4, 57]. It has gained wide application in various fields, including long term structural health monitoring (SHM) and assessment of infrastructure in civil engineering [30, 43, 74], oil and gas pipelines [38, 76], and geoscience [10, 35], just to name a few. To overcome the technical challenges of BOTDA systems such as spatial resolution, sensing range, measurement time etc., various strategies in the sensing interrogators have been proposed and implemented [3, 17, 23, 55]. Besides innovations in the sensing interrogators, an important aspect of research effort is placed and realized in the optical fiber cables for successful exploitation of Brillouin distributed sensing [6]. Various Brillouin scattering materials have been explored to provide reference and guidance to SBS related experiments and application development [2]. Integration of specialty fiber technology into BOTDA system leads to many exciting benefits in sensing performance improvement and opportunities for new sensing applications. There are several ways to explore and design the BOTDA system from the perspectives of fibers [20] or fiber cables [6]. Firstly, optimization through the optical and acoustic waveguide design of the fibers would result in novel fiber configurations that have potential for discriminative sensing; Secondly, the fiber materials, selections of dopants can enrich the design of the specialty fibers that can provide unconventional properties [20]; Lastly, sensing cable is another element that can be designed specifically for Brillouin sensing, including the selection of type, coating, materials, and geometry of the fiber cable component [6].

In this chapter, the focus is primarily on specialty fiber based BOTDA systems and applications. Other Brillouin sensing systems, such as Brillouin Optical Time Domain Reflectometry (BOTDR), Brillouin optical correlation-domain analysis (BOCDA) etc., based on specialty fibers, are not discussed within the scope of this chapter. In subsequent sections, the working principle of BOTDA is presented, followed by a brief overview of specialty fiber technologies that are applied in

BOTDA systems. Subsequently, the examples of the specialty optical fiber enabled BOTDA systems are presented, including optical fibers with special materials, special structures, and multiplexed fiber sensing systems. An overview of applications enabled by specialty fiber based BOTDA is presented. Lastly, the prospects of using specialty fibers for BOTDA systems are discussed.

2 BOTDA Technology and Applications

In a BOTDA system, the Brillouin interaction between a pulsed pump and a counter-propagating continuous wave (CW) probe can be measured by the schematic setup in Fig. 1. The frequency difference between the pulsed and the CW light is turned to the Brillouin frequency of the fiber associated with amplification of the probe light [11]. The probe power is measured as a function of time, which is translated to a distance-dependent information to derive local Brillouin gain. The local BFS at each point of the fiber is derived from reconstruction of Brillouin gain spectrum (BGS) by a frequency sweep with every frequency detuning around the Brillouin frequency.

Specialty fibers such as photonic crystal fibers (PCFs) and hollow core fibers are fabricated using silica and other non-silica materials, which introduce various design flexibility in applications [41]. Their holey structure is another dimension for waveguide design and properties optimization for sensing performance. Combining the material and the structural tunability, fibers provide an ideal platform to integrate with high gain SBS medium to boost the overall Brillouin amplification [71, 72], and/or achieve peculiar functions such as polarization noise suppression [8], enhancement of sensitivity through suitable coatings [29, 44] etc. Distributed sensing systems such as BOTDA system face crucial challenges in distinguishing between temperature and strain responses. This chapter examines the strategies employed to address this challenge, including the incorporation of specialty fibers such as photonic crystal fibers (PCFs), few mode fibers (FMFs), and multicore fibers into the BOTDA solution. The review delves into the introduction of multiplexed distributed sensing,

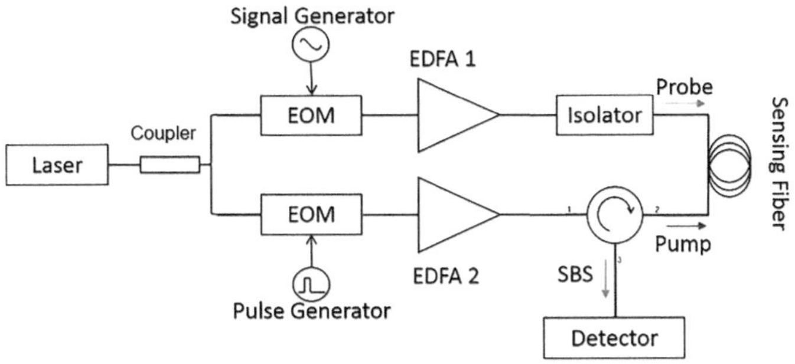

Fig. 1 Schematic setup of BOTDA system (Reprinted/Adapted] with permission from [34])

combining BOTDA technology with other Distributed Fiber Sensing (DFS) technologies like Raman optical time domain reflectometry (R-OTDR) or phase-sensitive optical time-domain reflectometry (φ-OTDR) through the utilization of multi-core fibers and spatial multiplexing. These advancements and opportunities have sparked significant interest, propelling the evolution of specialty fiber based BOTDA systems with notable features and performance improvements, as outlined in the subsequent section providing an overview of specialty fiber technology in BOTDA systems.

BOTDAs with optical fibers of special material: Non-silica materials such as polymer are explored to develop advanced fiber sensors, utilizing their unique properties such as high flexibility, large breakage strain and impact resistance, thus enabling applications that demand large sensitivities, large scale strain sensing etc. [54]. Besides through waveguide design, the Brillouin scattering characteristics such as frequency shift, spectral linewidth, and Brillouin gain can be widely tuned through intrinsic glass materials [13].

Compared to silica optical fibers which have a fundamental limitation in flexibilities and often break at relatively small strains of <2%–3%, polymer or plastic optical fibers are attracting great attention to overcome this fundamental issue. The BFS and the Brillouin bandwidth in a perfluorinated graded-index (PFGI) polymer optical fiber (POF) with 120 μm core diameter were characterized as 2.83 GHz and 105 MHz [52], showing the potential of the utilization of more economical Photodetectors (PD) and Electronic Spectrum Analyzers (ESA) with reduced bandwidth requirements as compared to standard silica fibers. Furthermore, PFGI Polymer fibers demonstrate better mode coupling efficiency and excellent mode scrambling capability regardless of alignment conditions [19]. The temperature and strain sensitivity of the PFGI-POF were experimentally characterized, showing negative values which are -3.5 and -0.2 times as compared to silica fibers for strain-insensitive high accuracy temperature sensing [53]. However, temperature and strain compensation are necessary for accurate POF-based Brillouin sensing due the cross effect of strain and temperature coefficients in selected strain ranges and temperature range of 32–60 °C. Firstly, in the strain range of 0–1.2%, the temperature coefficient has a linear relationship with strain, with a coefficient of 1.5 MHz/(%·°C). Secondly, in the strain ranges of 0–1.2% and 4.0–9.0%, the corresponding temperature dependencies are linear, with coefficients of 1.5 and −0.3 MHz/(%·°C), respectively. Thirdly, for strains exceeding 13%, temperature sensing without sensitivity to strain becomes a potential option using BFS in POFs [46]. When large strains are applied on PFGI-POF, the Stokes power drastically drops when the applied strain was larger than ~10%. Meanwhile, the dependance of BFS is highly non-monotonic, posing difficulties in accurately determining strain solely through BFS measurements. For strain range 10–15%, the BFS in PFGI-POF has no strain dependance, making it ideal and practical for temperature sensing [32]. Another study of BFS hopping phenomenon revealed that BFS abruptly changes from 2.7 GHz to 3.2 GHz due to the strain-induced slimming of the over-cladding layer composed of polycarbonate, triggered by the elastic-to-plastic transition [31]. Wide dynamic range of strain measurement >25% has been achieved by applying 3.0% strain to a slimmed POF beforehand [50].

The BFS dependence on water absorption ratio has been observed in standard fibers, showing nonlinear displacement with hysteretic behavior. Thus, the humidity isolation is a critical factor for achieving precise BFS measurements [27]. In PMMA fibers, the BFS decreased monotonically as the water absorption ratio increased, due to decreasing Young's modulus. Moreover, at the same water absorption ratio, the BFS change was larger at a higher temperature [45]. The humidity induced BFS in PFGI-POF has been characterized, showing very different characteristics from standard fibers. Firstly, there is a linear relationship between BFS and humidity at constant temperatures. Secondly, there is no hysteresis observed due to humidity changes. Thirdly, the temperature coefficient of BFS is independent of humidity variations. Lastly, an analytical description has been proposed to decouple the temperature and humidity as two mutually independent effects on the BFS, enabling a quantitative estimation of BFS for specified temperature and humidity [61]. The humidity induced BFS was predominantly caused by fiber straining because of swelling of the fiber over-cladding. In addition, PFGI-POFs demonstrate a greater humidity-induced strain response compared to standard silica fibers coated with polyimide and acrylate. The measured strain coefficient of (-146.5 ± 0.9) MHz/% within strain range from 0.1% to 1.5%. The fiber expansion coefficient for humidity and temperatures are $(7.4 \pm 0.1) \bullet 10^{-6}$%r.h.$^{-1}$ and $(22.7 \pm 0.3) \bullet 10^{-6}$ K^{-1} [62]. The potential of using polymer optical for distributed pressure sensing is reported [51], showing a significant pressure sensitivity 4.3 MHz/MPa which is 5.8 times of bare silica fibers with opposite sign. In addition, the hysteresis of the pressure response is mitigated after several cycles of increasing/decreasing pressure.

Tellurite fibers with large Brillouin gain coefficients of 1.7×10^{-10} m/W in BOCDR configuration have been reported to show negative strain coefficient − 230 MHz/% and temperature coefficient −1.14 MHz/ °C of the BFS, respectively [48, 49].

Bismuth-oxide nonlinear fibers were demonstrated to have superior Kerr nonlinear efficiency over silica fibers [39]. The temperature coefficient of the BFS in Bismuth-oxide nonlinear fibers were measured by BOCDR system, with the value of -0.88 MHz/°C [48].

Rare-earth-doped fibers, such as Erbium-doped fibers (EDFs) with different erbium concentrations of 0.71, 1.2 and 2.28 wtppt and the same cladding, were investigated by Ding et al. [15]

Strong SBS in single mode chalcogenide glass fibers were observed, demonstrating Brillouin gain coefficient one to two orders of magnitude larger than that of fused silica fiber [1, 24, 25]. For example, the estimated Brillouin gain coefficient was $(3.9 \pm 0.4) \times 10^{-9}$ m/W for the As_2S_3 and $(6.75 \pm 0.35) \times 10^{-9}$ m/W for As_2Se_3 [24], 8×10^{-10} m/W for Ge15Sb20S_{65} chalcogenide glass microstructured fiber made of germanium sulfide glass [25]. Slow and fast light generation based on SBS was demonstrated in As2Se3 chalcogenide fiber, showing Brillouin gain of 43 dB with 60 mW pump power in a 5 m chalcogenide fiber, which results in 37 ns optical delay of a probe Gaussian pulse of 50 ns duration [64]. Hybrid PMMA- As_2Se_3 microwires demonstrated significant increase in the Brillouin threshold as compared to the As_2Se_3 fibers due to acoustic damping in the PMMA cladding. In addition, hybrid acoustic modes with both shear and longitudinal phonons can be excited in the microwire. The BFS can be widely tuned by changing the microwire diameters [7]. Strong coupling of forward and backward Brillouin scattering between core and cladding in a 60 cm hybrid As2Se3-PMMA microtapers was observed, generating amplification of Stokes wave at 7.4 GHz due to modulation by guided-acoustic-wave Brillouin scattering, and two peaks from the interaction of the pump with the Stokes wave through SBS at 7.62 GHz and 7.8 GHz [59]. Simultaneous temperature and strain sensing based on a tapered dual-core As2Se3-PMMA fiber was demonstrated through different BFS response in two principal polarization axes. The temperature coefficients are -3.8272 MHz/°C and −3.2202 MHz/°C, and the strain coefficients are -614.3 MHz/% and -346.3 MHz/% [66]. Due to low Young's modulus of the As2Se3 core and the high mechanical strength of PMMA cladding, wide dynamic strain range from 0 to 15,000 $\mu\varepsilon$ based on Brillouin frequency and linewidth in an 50 cm-long As2Se3-PMMA hybrid microfiber was achieved. No strain memory effect was observed. This study provided compelling evidence and highlighted the potential of using As2Se3-PMMA hybrid microfiber for wide range strain sensing applications [67]. The birefringence influence on SBS was investigated in elliptical-core As2Se3-PMMA microfibers. For a fixed ellipticity of 1.108, the decrease in the minor axis of the core diameter from 1.50 μm to 0.87 μm led to increasing group birefringence from ~10^{-3} to ~10^{-2}, and larger BFS difference from ~6 MHz to ~30 MHz, associated with broadened BGS from ~70 MHz to ~140 MHz [66]. Distributed temperature and strain sensing based on a nonuniform As2Se3-PMMA tapered fiber with spatial resolution of 1.1 cm was demonstrated by Brillouin gain-profile tracing, using differential BGS at each fiber location [28]. Saxena et al. reported the calibration of BFS induced by a transverse load on a 60 cm single-mode As2Se3-PMMA microtaper, showing sensitivity of 0.08 MHz/N [60].

The comparisons of strain and temperature sensitivities of the BFS among different fiber materials, such as silica fibers, non-silica fiber materials including polymer, Tellurite, Bismuth-oxide, Erbium-doped, and As2Se3-PMMA fibers, are listed in Table 1.

Table 1 Comparison of strain and temperature sensitivities of the BFS among different fiber materials

	Strain sensitivity (MHz/%)	Temperature sensitivity (MHz/°C)	Selected reference(s)
Silica fibers	580	1.18	[33, 37]
Polymer fibers	−121.8	−4.09	[53]
Tellurite fibers	−230	−1.14	[49]
Bismuth-oxide fibers	NA	−0.88	[48]
Erbium-doped fibers	479	0.87	[15]
Dual core As2Se3-PMMA fiber tapers	−614.3 −346.3	−3.8272 −3.3302	[68]

BOTDAs with optical fibers of special structure: Optical fibers with special structures bring new features and flexibilities in controlling and optimizing the optical and acoustic waveguide properties, thus enabling new applications, and enhancing sensing performance in BOTDA systems. To realize discriminative sensing of temperature and strain in BOTDA systems, there are two main approaches, primarily by constructing a sensor matrix of temperature and strain coefficients in the sensing system and based on a combination of using BFS sensing and other parameters sensing.

The sensor matrix is typically formulated as Eq. 1, in which $C_{T1,2}$ and $C_{\varepsilon1,2}$ represent the temperature and strain coefficients, $\Delta\nu_{B\_peak1,2}$ represent the central frequency shifts of the Brillouin spectra peaks. The temperature change ΔT and the strain change $\Delta\varepsilon$ along the fiber are calculated by solving the inverse matrix problem.

$$\begin{bmatrix} \Delta\nu_{B\_peak1} \\ \Delta\nu_{B\_peak2} \end{bmatrix} = \begin{bmatrix} C_{T1} & C_{\varepsilon1} \\ C_{T2} & C_{\varepsilon2} \end{bmatrix} \begin{bmatrix} \Delta T \\ \Delta\varepsilon \end{bmatrix} \tag{1}$$

The reported approaches to achieve discrimination of temperature and strain by specialty fiber are summarized in Table 2 [34], including exploiting the multi-peak BGS due to multiple acoustic modes and/or multiple optical modes, multiple core/fibers, BFS and other parameters such as birefringence, fluorescence, and athermal/atensic optical fibers.

The multiple peaks in the BGS observed in specialty fibers exhibit different strain and temperature sensitivities, which enables the construction of the sensor matrix for discriminative sensing. The first measurement of dependence of strain and temperature in PCF was reported in [80]. The principles of producing multiple peaks in the BGS are summarized below. Firstly, the multiple peaks of BGS are due to the non-zero overlapping between multiple acoustic modes and the fundamental optical mode, as reported in PCFs with a partially germanium-doped core [79, 80], and in large effective-area fiber (LEAF) [42]. Secondly, the interaction between multiple optical modes and the fundamental acoustic modes results in multiple peaks in the BGS as well, as observed in [40]. Thirdly, choosing proper optical-acoustic mode pairs in an elliptical-core FMF [22] from the interaction between higher order optical modes and higher order acoustic modes can be another alternative approach to optimize the sensing performance of temperature and strain discrimination in BOTDA.

Combination of BFS sensing and other parameter sensing is another effective approach. Polarization-maintaining photonic crystal fiber (PM-PCF) was employed in the sensing system to perform monitoring of BFS and birefringence simultaneously and reported discrimination accuracy of 3 $\mu\varepsilon$ and 0.08 °C [82]. Combining BFS and fluorescence measurement was reported in erbium-doped optical fiber-based sensing system to demonstrate strain and temperature discriminative sensing [16]. Lastly, temperature-free strain sensing or strain-free temperature sensing could be possible if the sensing fibers are immune to one parameter only. As reported in [14], highly Ge doped fibers exhibit negligible temperature sensitivity if the fiber coating is removed, presenting a good potential for single parameter sensing. However, the attempts to

Table 2 Summary of reported approaches to achieve discriminative sensing using specialty fibers (Reprinted/Adapted] with permission from [34])

Approaches	Specialty fiber type	Temperature coefficient $C_T(MHz/°C)$	Strain coefficient $C_\varepsilon(MHz/\mu\varepsilon)$
Multi-peak BGS due to multiple acoustic modes	PCFs with a partially germanium-doped core [79, 81] Peak 1 Peak 2	0.96 1.25	0.048 0.055
	Large effective-area fiber (LEAF) [42] Peak 1 BFS Peak 1 BGS linewidth Peak 2 BFS Peak 2 BGS linewidth	1.18 −0.111 1.2 −0.118	0.055 −0.00059 0.055 −0.00091
Multi-peak BGS due to multiple optical modes	Circular core-few mode fiber (FMF) [40] LP_{01} LP_{11}	1.0169 0.99099	0.05924 0.04872
Multi-peak BGS due to multiple acoustic modes & optical modes	Elliptical core FMF [22] LP_{01}-P_1 LP_{01}-P_2 LP_{11e}-P_1 LP_{11e}-P_2	1.242 1.278 1.287 1.501	0.0613 0.0364 0.0658 0.0484
Multiple core/fibers	Different fiber deployment [5]	1.2±0.2	0.054±0.7
	Dual core fiber [73] Core 1 Core 2	0.971 0.9593	0.0532 0.0729
	Multicore fiber (MCF) [47] Central core Outer core MCF with heterogeneous cores [77] Central core Outer core	1.08 1.03 1.05 ± 0.095 1.15±0.0343	0.0485 0.0517 0.0485 0.0485
BFS and birefringence	Polarization-maintaining photonic crystal fiber (PM-PCF) [75] Birefringence BFS	−40°C to −15°C: 27.4 −15°C to 5°C: 8 5°C to 80°C: 0 1.15	−0.154 0.049
BFS and fluorescence	Erbium-doped optical fiber [16] Fluorescence intensity ratio BFS	5.6×10^{-4}/°C 0.87	0 0.0479
Athermal/atensic optical fibers	Highly Ge dope fibers [14]	0.07	0.0214

create athermal or atensic fibers by optimizing the composition of fiber materials encounter a substantial decrease in sensitivity to strain or temperature, as well as in Brillouin gain [20].

Multiplexed distributed fiber sensing system: Firstly, hybrid Brillouin-Rayleigh sensing systems are attracting attention as they are capable of dual parameter measurements for strain and temperature.

In this hybrid system, standard single mode fibers can be used as sensing fiber. Hybrid Brillouin-Rayleigh sensing system was also reported to overcome the challenges of complex interplay among scattering effects inside fiber, the interaction of two sensing signals, and the competition of pump power. Frequency division multiplexing (FDM), wavelength division multiplexing (WDM), and time division multiplexing (TDM) are fused in the system to achieve ultra-long sensing range of 150.62 km, spatial resolution of 9 m by BOTDA, and 30 m by Φ-OTDR respectively. As depicted in Fig. 2, the Φ-OTDR has little impact on the BFS sensing results of BOTDA. Likewise, the Rayleigh scattering signals in the system showed little difference by comparing the intensity trace with and without Brillouin pump, and with/without BOTDA [26]. To achieve compact hybrid Brillouin-Rayleigh sensing system, multicore fibers provide spatially separated sensing channels based on the central core and an outer core within the same fiber, for BOTDA and Φ-OTDR, respectively. The hybrid sensing system achieved a good combination of large measurement range of 10 °C by BOTDA, and small resolution of ~0.001 °C by Φ-OTDR [12]. This combination represents a promising approach to tackle such trade-off challenges in the sensing system.

Secondly, hybrid Brillouin-Raman sensing systems are developed for distributed sensing of temperature and strain, employing promising techniques based on pulse coding in mitigating the challenges associated with the tradeoff between sensing distance and spatial resolution [56].

Selected applications by specialty fiber enabled BOTDA sensing: The BFS difference produced by the central core and outer core presents a desirable opportunity of using multicore fibers for curvature sensing and shape reconstruction [78]. Ring core fibers are another candidate for curvature sensing. Being more bend resistant than SMF, RCF enables distributed curvature sensing with BFS ~32.9 MHz for a radius of 0.5 cm, without temperature or axial strain sensitivity by measuring its Brillouin gain variation [63]. Optical fibers with double layer polymer coatings are used to enhance the hydrostatic pressure sensitivity and demonstrated for sensitivity-enhanced distributed hydrostatic pressure sensing based on BOTDA. Four kinds of fibers, including single layer coated, double layer coated fiber with outer layer radius of 450, 1000, and 1500 μm were experimentally demonstrated to achieve BFS pressure sensitivity of −0.75 MHz/MPa, −1.65 MHz/MPa, − 2.66 MHz/MPa and − 3.61 MHz/MPa in the range of 0–30 MPa after temperature compensation, respectively [18]. Dynamic pressure measurement based on frequency-agile fast BOTDA and the sensing fiber with double layer coating was demonstrated with promising sensing performance, including static pressure sensitivity of −3.46 MHz/MPa in the range of 0–30 MPa, and large dynamic pressure range with sampling rate of

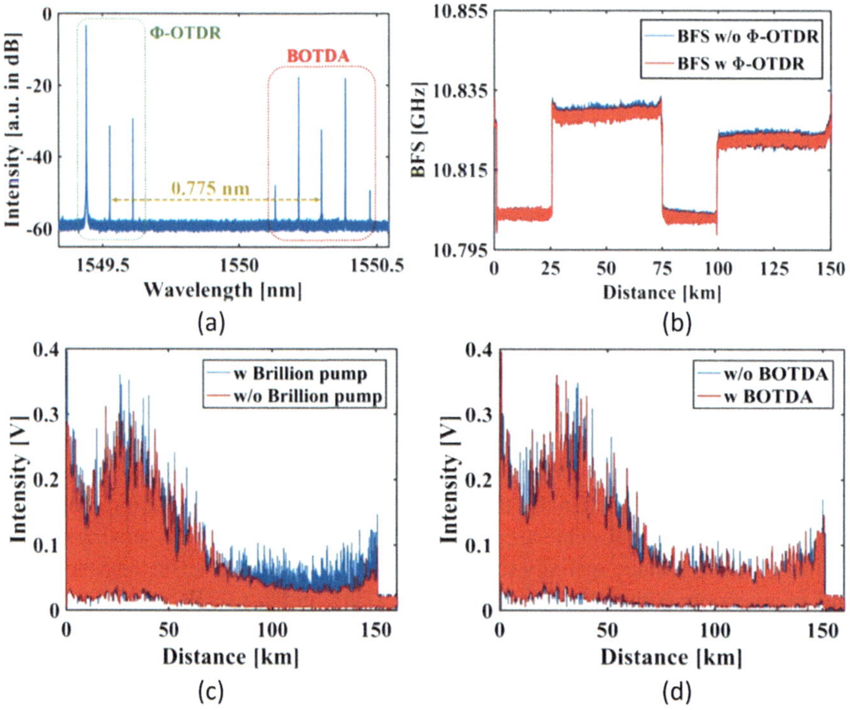

Fig. 2 **a** Optical spectrum of the hybrid sensing signals; **b** BFS with/without the Φ-OTDR; **c** Φ-OTDR intensity traces with/without the Brillouin pump; **d** Φ-OTDR intensity traces the operation of BOTDA (Reprinted/Adapted] with permission from [26])

33.3 kHz, maximum sensing distance of 25 m, and uncertainty of dynamic pressure measurement system is 0.03 MPa [58]. GeO2-doped SMF and pure silica PCF based BOTDA systems for high temperature sensing up to 1100 °C and 1200 °C are demonstrated, showing measurement accuracy as high as ±2.4 °C for the SMF and ±3.6 °C for the PCF [70]. An annealed single mode gold-coated optical fiber was tested for temperature and strain measurement at high temperatures up to 1000 °C as shown in Fig. 3. The strain coefficient changed from ~0.046 MHz/με at 100 °C to ~0.022 MHz/με at 1000 °C, mainly due to the increase in Young's modulus of silica with temperature [69]. Potential applications of distributed high-temperature strain sensing include asset structural integrity monitoring based on strain monitoring in high temperature environments such as power plants and jet engines.

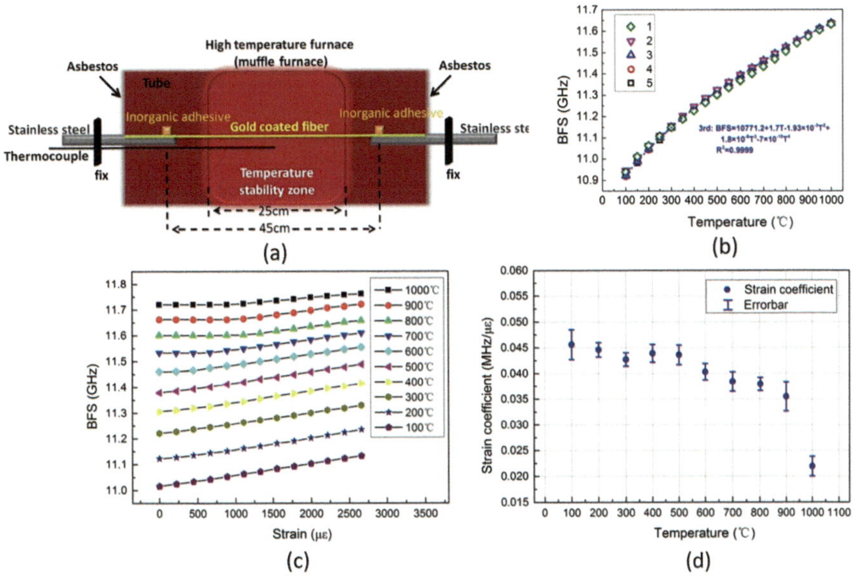

Fig. 3 **a** Experimental setup of gold-coated fiber for high-temperature strain measurement with a spatial resolution of 5 cm. **b** 5 measurements of a single sensing point from 100 to 1000 °C with a step of 50 °C, temperature coefficient of gold-coated fiber changes with temperature nonlinearly; High-temperature strain coefficients of annealed gold-coated fiber from 100 to 1000 °C. **c** BFS with strain at a position of 3.6 m; **d** strain coefficients for all the data in a span of 25 cm [Reprinted/Adapted] with permission from [69] © Optical Society of America

3 Outlook and Conclusion

This chapter highlights various achievements utilizing specialty fibers in BOTDA systems, covering key aspects such as the unique properties of large sensitivities, large scale strain introduced by special material into the sensing fiber, sensitivity improvements through specially coated optical fibers, discriminative sensing of temperature and strain, and realization of novel sensing parameters enabled by specialty fibers. It is increasingly recognized that specialty fibers play a crucial role in the design and implementation of BOTDA systems for distributed sensing applications. Additionally, the ongoing development of specialty fibers tailored for sensing applications, achieved through optimization of topology (core shape, number of modes, multi-cores) and the incorporation of material properties (silica/air, gas, highly Ge-doped), such as multi-material fibers, holds the promise of opening new avenues for further enhancing BOTDA system performance.

References

1. Abedin KS. Observation of strong stimulated Brillouin scattering in single-mode As_2Se_3 chalcogenide fiber. Opt Express. 2005;13(25):10266–71.
2. Bai Z, et al. Stimulated Brillouin scattering materials, experimental design and applications: a review. Opt Mater. 2018;75:626–45.
3. Bao X, Chen L. Recent progress in brillouin scattering based fiber sensors. Sensors. 2011;11:4252–87.
4. Bao X, Webb DJ, Jackson DA. 32-km distributed temperature sensor based on Brillouin loss in an optical fiber. Opt Lett. 1993;18(18):1561–3.
5. Bao X, Webb DJ, Jackson DA. Combined distributed temperature and strain sensor based on Brillouin loss in an optical fiber. Opt Lett. 1994;19(2):141–3.
6. Bastianini F, et al. Optical fiber sensing cables for brillouin-based distributed measurements. Sensors. 2019;19(23):5172.
7. Beugnot J-C, et al. Reduction and control of stimulated Brillouin scattering in polymer-coated chalcogenide optical microwires. Opt Lett. 2014;39(3):482–5.
8. Beugnot J-C, et al. Complete experimental characterization of stimulated Brillouin scattering in photonic crystal fiber. Opt Express. 2007;15(23):15517–22.
9. Brillouin L. Diffusion de la lumière et des rayons X par un corps transparent homogène. Ann Phys. 1922;9(17):88–122.
10. Cao D-F, et al. Characterization of soil moisture distribution and movement under the influence of watering-dewatering using AHFO and BOTDA technologies. Environ Eng Geosci. 2019;25(3):189–202.
11. Chiao RY, Townes CH, Stoicheff BP. Stimulated Brillouin scattering and coherent generation of intense supersonic waves. Phys Rev Lett. 1964;12(21):592–5.
12. Dang Y, et al. Towards large dynamic range and ultrahigh measurement resolution in distributed fiber measurement resolution in distributed fiber. Opt Express. 2017;25(17):20183–93.
13. Deroh M, et al. Comparative analysis of stimulated Brillouin scattering at 2-μm in various infrared glass-based optical fibers. J Opt Soc America B. 2020;37(12):3792–800.
14. Deroh M, et al. Towards athermal Brillouin strain sensing based on heavily germania-doped core optical fibers. APL Photonics. 2019;4: 030801.
15. Ding M, Hayashi N, Mizuno Y, Nakamura K. Brillouin gain spectrum dependences on temperature and strain in erbium-doped optical fibers with different erbium concentrations. Appl Phys Lett. 2013;102(19): 191906.
16. Ding M, Mizuno Y, Nakamura K. Discriminative strain and temperature measurement using Brillouin scattering and fluo-rescence in erbium-doped optical fiber. Opt Express. 2014;22(20):24706–12.
17. Dong Y. High-performance distributed brillouin optical fiber sensing. Photon Sens. 2021;11(1):69–90.
18. Dong Y, et al. Sensitivity-enhanced distributed hydrostatic pressure sensor based on BOTDA in single-mode fiber with double-layer polymer coatings. J Lightwave Technol. 2020;38(8):2564–71.
19. Dong Y, et al. Characterization of evolution of mode coupling in a graded-index polymer optical fiber by using Brillouin optical time-domain analysis. Opt Express. 2014;22(22):26510–6.
20. Dragic P, Ballato J. A brief review of specialty optical fibers for brillouin-scattering-based distributed sensors. Appl Sci. 2018;8(10):1996.
21. EP Ippen RHS. Stimulated Brillouin scattering in optical fibers. Appl Phys Lett. 1972;21:539.
22. Fang J, et al. Multi-parameter distributed fiber sensing with higher-order optical and acoustic modes. Opt Lett. 2019;44(5):1096–9.
23. Feng C, Kadum JE, Schneider T. The state-of-the-art of brillouin distributed fiber sensing. In: Fiber optic sensing—principle, measurement and applications. s.l.:intechopen, p. 84684 (2019).
24. Florea C, et al. Stimulated Brillouin scattering in single-mode As_2S_3 and As_2Se_3 chalcogenide fibers. Opt Express. 2006;14(25):12063–70.

25. Fortier C, et al. Experimental investigation of Brillouin and Raman scattering in a 2SG sulfide glass microstructured chalcogenide fiber. Opt Express. 2008;16(13):9398–404.
26. Fu Y, et al. Ultra-long-distance hybrid BOTDA/Φ-OTDR. Sensors. 2018;18:976.
27. Galindez C, Madruga FJ, Lopez-Higuera JM. Influence of humidity on the measurement of brillouin frequency shift. IEEE Photonics Technol Lett. 2008;20(23):1959–61.
28. Gao S, et al. Monitoring local temperature and longitudinal strain along a nonuniform As2Se3-PMMA tapered fiber by Brillouin gain-profile tracing. Opt Express. 2022;30(16):29655–64.
29. Gu H, et al. Effects of polymer coatings on temperature sensitivity of brillouin frequency shift within double-coated fibers. IEEE Sens J. 2013;13(2):864–9.
30. Guo T, et al. Experimental study on strain and deformation monitoring of reinforced concrete structures using PPP-BOTDA. Sci China Ser E: Technol Sci. 2009;59:2859–68.
31. Hayashi N, Minakawa K, Mizuno Y, Nakamura K. Brillouin frequency shift hopping in polymer optical fiber. Appl Phys Lett. 2014;105(9): 091113.
32. Hayashi N, Mizuno Y, Nakamura K. Brillouin gain spectrum dependence on large strain in perfluorinated graded-index polymer optical fiber. Opt Express. 2012;20(19):21101–6.
33. Horiguchi T, Kurashima T, Tateda M. Tensile strain dependence of Brillouin frequency shift in silica optical fibers. IEEE Photonics Technol Lett. 1989;1(5):107–8.
34. Hu DJJ, et al. Review of specialty fiber based Brillouin optical time domain analysis technology. Photonics. 2021;8(10):421.
35. Iten M, Hauswirth D, Puzrin A. Distributed fiber optic sensor development, testing, and evaluation for geotechnical monitoring applications. San Diego, California, United States, s.n. (2011).
36. Kobyakov A, Sauer M, Chowdhury D. Stimulated Brillouin scattering in optical fibers. Advances in Optics and Photonics. 2010;2(1):1–59.
37. Kurashima T, Horiguchi T, Tateda M. Thermal effects of Brillouin gain spectra in single-mode fibers. IEEE Photonics Technol Lett. 1990;2(10):718–20.
38. Kwon I-B et al. Feasibility study for monitoring of offshore pipelines using BOTDA system. Orlando, Florida, United States, s.n. (2009).
39. Lee JH, et al. Experimental comparison of a Kerr nonlinearity figure of merit including the stimulated Brillouin scattering threshold for state-of-the-art nonlinear optical fibers. Opt Lett. 2005;30(13):1698–700.
40. Li A, et al. Few-mode fiber multi-parameter sensor with distributed temperature and strain discrimination. Opt Lett. 2015;40(7):1488–91.
41. Liu H, et al. Specialty optical fibers for advanced sensing applications. Opto-Electron Sci. 2023;2: 220025.
42. Liu X, Bao X. Brillouin spectrum in leaf and simultaneous temperature and strain measurement. J Lightwave Technol. 2012;30(8):1053–9.
43. Liu Y, et al. Damage detection of tunnel based on the high-density cross-sectional curvature obtained using strain data from BOTDA sensors. Mech Syst Signal Process. 2021;158: 107728.
44. Lu X, Soto MA, Thévenaz L. Impact of the fiber coating on the temperature response of distributed optical fiber sensors at cryogenic ranges. J Lightwave Technol. 2017;36(4):961–7.
45. Minakawa K, et al. Dependence of Brillouin frequency shift on water absorption ratio in polymer optical fibers. J Appl Phys. 2016;119(22): 223102.
46. Minakawa K, Mizuno Y, Nakamura K. Cross effect of strain and temperature on brillouin frequency shift in polymer optical fibers. J Lightwave Technol. 2017;35(12):2481–6.
47. Mizuno Y, et al. Brillouin scattering in multi-core optical fibers for sensing applications. Sci Rep. 2015;5:11388.
48. Mizuno Y, He Z, Hotate K. Dependence of the Brillouin frequency shift on temperature in a tellurite glass fiber and a bismuth-oxide highly-nonlinear fiber. Appl Phys Express. 2009;2(11): 112402.
49. Mizuno Y, He Z, Hotate K. Distributed strain measurement using a tellurite glass fiber with Brillouin optical correlation-domain reflectometry. Optics Commun. 2010;283(11):2438–41.
50. Mizuno Y, Lee H, Hayashi N, Nakamura K. Hydrostatic pressure dependence of Brillouin frequency shift in polymer optical fibers. Appl Phys Express. 2018;11(1): 012502.

51. Mizuno Y, et al. Brillouin characterization of slimmed polymer optical fibers for strain sensing with extremely wide dynamic range. Opt Express. 2018;26(21):28030–7.
52. Mizuno Y, Nakamura K. Experimental study of Brillouin scattering in perfluorinated polymer optical fiber at telecommunication wavelength. Appl Phys Lett. 2010;97(2): 021103.
53. Mizuno Y, Nakamura K. Potential of Brillouin scattering in polymer optical fiber for strain-insensitive high-accuracy temperature sensing. Opt Lett. 2010;35(23):3982–7.
54. Mizuno Y, et al. Distributed polymer optical fiber sensors: a review and outlook. Photonics Res. 2021;9(9):1719–33.
55. Motil A, Bergman A, Tur M. State of the art of Brillouin fiber-optic distributed sensing. Opt Laser Technol. 2016;78:81–103.
56. Muanenda Y, Oton CJ, Pasquale FD. Application of raman and brillouin scattering phenomena in distributed optical fiber sensing. Frontiers in Physics. 2019;7:155.
57. Niklès M, Thévenaz L, Robert PA. Simple distributed fiber sensor based on Brillouin gain spectrum analysis. Opt Lett. 1996;21(10):758–60.
58. Qiu L, et al. Dynamic distributed pressure measurement using frequency-agile fast BOTDA. IEEE Sens J. 2021;21(22):25743–8.
59. Saxena B, Baker C, Bao X, Chen L. High birefringent Brillouin frequency shifts in a single-mode As_2Se_3-PMMA microtaper induced by a transverse load. Opt Lett. 2019;44(19):4789–92.
60. Saxena B, Baker C, Bao X, Chen L. Simultaneous generation of guided-acoustic-wave Brillouin scattering and stimulated-Brillouin-scattering in hybrid As_2Se_3-PMMA microtapers. Opt Express. 2019;27(10):13734–43.
61. Schreier A, Liehr S, Wosniok A, Krebber K. Investigation on the influence of humidity on stimulated brillouin backscattering in perfluorinated polymer optical fibers. Sensors. 2018;18(11):3952.
62. Schreier A, Wosniok A, Liehr S, Krebber K. Humidity-induced Brillouin frequency shift in perfluorinated polymer optical fibers. Opt Express. 2018;26(17):22307–14.
63. Shen L, et al. Distributed curvature sensing based on a bending loss-resistant ring-core fiber. Photonics Research. 2020;8(2):165–74.
64. Song KY, et al. Highly efficient Brillouin slow and fast light using As_2Se_3 chalcogenide fiber. Opt Express. 2006;14(13):5860–5.
65. T Horiguchi, Kurashima T, Tateda M. A technique to measure distributed strain in optical fibers. IEEE Photonics Technol Lett. 1990;2(5):352–354.
66. Wang H, Baker C, Chen L, Bao X. Stimulated Brillouin scattering in high-birefringence elliptical-core As2Se3-PMMA microfibers. Opt Lett. 2021;46(5):945–8.
67. Wang H, et al. Stimulated Brillouin scattering in a tapered dual-core As_2Se_3-PMMA fiber for simultaneous temperature and strain sensing. Opt Lett. 2020;45(12):3301–4.
68. Wang H, et al. Wide-range strain sensor based on Brillouin frequency and linewidth in an As_2Se_3-PMMA hybrid microfiber. Opt Express. 2020;28(15):22933–45.
69. Xu P, et al. Distributed Brillouin optical fiber temperature and strain sensing at a high temperature up to 1000 °C by using an annealed gold-coated fiber. Opt Express. 2018;26(23):29724–34.
70. Xu P, et al. 1200 °C high-temperature distributed optical fiber sensing using Brillouin optical time domain analysis. Appl Opt. 2016;55(21):5471–8.
71. Yang F, Gyger F, Thévenaz L. Giant Brillouin amplification in gases. Opt Photonics News. 2020;31(12):33–33.
72. Yang F, Gyger F, Thévenaz L. Intense Brillouin amplification in gas using hollow-core waveguides. Nat Photonics. 2020;14:700–8.
73. Zaghloul MA, et al. Discrimination of temperature and strain in brillouin optical time domain analysis using a multicore optical fiber. Sensors. 2018;18(4):1176.
74. Zhang D, et al. Structural crack detection using DPP-BOTDA and crack-induced features of the brillouin gain spectrum. Sensors. 2020;20(23):6947.
75. Zhang H, et al. Simultaneous measurement of strain and temperature using a polarization-maintaining photonic crystal fiber with stimulated Brillouin scattering. Appl Phys Express. 2017;10: 012501.

76. Zhang S, Liu B, He J. Pipeline deformation monitoring using distributed fiber optical sensor. Measurement. 2019;133:208–13.
77. Zhao Z, et al. Spatial-division multiplexed Brillouin distributed sensing based on a heterogeneous multicore fiber. Opt Lett. 2017;42(1):171–4.
78. Zhao Z, Soto MA, Tang M, Thévenaz L. Distributed shape sensing using Brillouin scattering in multi-core fibers. Opt Express. 2016;24(22):25211–23.
79. Zou L, Bao X, Chen L. Brillouin scattering spectrum in photonic crystal fiber with a partially germanium-doped core. Opt Lett. 2003;28(21):2022–4.
80. Zou L, Bao XV, SA, Chen L. Dependence of the brillouin frequency shift on strain and temperature in a photonic crystal fiber. Optics Letters. 2004;29(13):1485–1487.
81. Zou L, Bao XVSA, Chen L. Dependence of the Brillouin frequency shift on strain and temperature in a photonic crystal fiber. Opt Lett. 2004;29(13):1485–1487.
82. Zou W, He Z, Hotate K. Complete discrimination of strain and temperature using Brillouin frequency shift and birefringence in a polarization-maintaining fiber. Opt Express. 2009;17(3):1248–55.

Chapter 3
In-Fiber Mach–Zehnder Interferometers for Sensing

Shiying Xiao, Beilei Wu, Shaoyang Ma, and Tao Ye

Abstract In-fiber Mach–Zehnder interferometers (MZIs) are based on the interference between two light beams transmitted in the functional optical fiber. They have the advantages of compact structure, high sensitivity, resistance to electromagnetic interference and most importantly, the flexibility to construct with different kinds of fibers and materials. Several methods, including the splicing method, the femtosecond laser inscribing method, the microtapering method and the macrobending method can be utilized to fabricate the in-fiber MZIs. The various advantages as well as different manufacturing methods allow the in-fiber MZIs to flourish in the sensing application areas. They are applied for the measurement of temperature, strain, curvature, refractive index (RI) as well as other measurands such as humidity, magnetic field and the concentration of gas and metal ions. In this chapter, the basic principle and sensing mechanism of the MZIs for the measurement of these measurands are presented and discussed. In addition, recent progress of in-fiber MZIs for these sensing applications is reviewed. Specifically, the means to enhance the sensitivities are summarized and discussed. The chapter starts with the fundamental principle and fabrication methods of in-fiber MZIs. Then, the sensing mechanisms and experimental studies of the in-fiber MZIs proposed for different measurands are reviewed. Finally, a summary of

S. Xiao · B. Wu
Key Laboratory of All Optical Network and Advanced Telecommunication Network of EMC, Institute of Lightwave Technology, Beijing Jiaotong University, Beijing, China
e-mail: syxiao@bjtu.edu.cn

B. Wu
e-mail: wubeilei@bjtu.edu.cn

S. Ma (✉)
Office of Science and Technology, Northwestern Polytechnical University, 127 Youyi Road, Xi'an 710072, China
e-mail: mashaoyang@nwpu.edu.cn

T. Ye (✉)
Ministry of Education Key Laboratory of Micro/Nano Systems for Aerospace, Key Laboratory of Micro- and Nano-Electro-Mechanical Systems of Shaanxi Province, School of Mechanical Engineering, Northwestern Polytechnical University, Xi'an 710072, China
e-mail: yetao@nwpu.edu.cn

L. Wei (ed.), *Advanced Optical and Optoelectronic Fibers*, Advances in Optics and Optoelectronics, https://doi.org/10.1007/978-981-97-6218-7_3

the whole chapter and a prospect for the development of the in-fiber MZI sensors are given in the conclusion part.

1 Introduction

Mach–Zehnder interferometer was first proposed by physicists Ludwig Mach and Ludwig Zehnder in the 1890s. The original apparatus of MZI is a quite bulk configuration, consisting of several separate mirrors and beam splitters. The invention of MZI has triggered numerous research branches, one of which is the fiber-optic MZIs. A traditional fiber-optic MZI consists of two optical fiber couplers and two optical fiber arms. The two couplers are used for the splitting and combining of the transmitted light. The two optical fiber arms respectively serve as the sensing arm and the reference arm. In recent years, the rise of in-fiber MZIs has aroused the interest of many researchers. An in-fiber Mach–Zehnder interferometer consists of a lead-in fiber, a lead-out fiber and a functional fiber in-between. The joints between the lead-in/lead-out fiber and the functional fiber take the place of the conventional couplers and serve as the light splitter and combiner. In the meanwhile, the functional fiber replaces the two fiber arms in the traditional fiber-optic MZI and allows two light beams to propagate, typically the beam of the core mode and that of the cladding mode. Compared with the bulk configuration of traditional optical fiber MZIs, the structure of in-fiber MZIs is simpler, more stable, more compact and easier to align.

In-fiber MZIs have the advantages of compact structure, high sensitivity, resistance to electromagnetic interference and most importantly, the flexibility to construct with different kinds of fiber and materials. The merits allow them to flourish in the sensing applications. In the sensing applications, both common fibers, such as the commercial single mode fibers (SMFs) and multimode fibers (MMFs), and the specialty fibers, such as photonic crystal fibers (PCFs) and microfibers, are utilized as the functional fibers. Materials such as polymers, oxides and fluids are incorporated with the functional fibers to adapt to different sensing scenarios. For instance, polydimethylsiloxane (PDMS) is utilized to enhance the temperature sensitivity of microfiber-based MZI sensors [4, 36]. Graphene oxide (GO) is applied with a tapered thin core fiber (TCF) to achieve the sensing of ammonia concentration [9]. Recently, magnetic fluid is incorporated with a tapered SMF to construct an in-fiber MZI sensor for the sensing of magnetic field [10]. In this chapter, we give a comprehensive review on the recent progress of the reported in-fiber MZIs for sensing applications. Firstly, the basic principle of the in-fiber MZIs is presented. Secondly, the structures and the most commonly used fabrication methods of the in-fber MZIs are summarized. Thirdly, the sensing mechanism and the experimental studies of the in-fiber MZIs for the sensing of temperature, strain, curvature, RI and some other measurands such as humidity, magnetic field and concentration of harmful gases and metal ions are reviewed. Finally, the whole chapter is summarized and the future development of optical fiber MZI sensors is prospected.

2 Main Body Text

2.1 Basic Principle

Figure 1a depicts an in-fiber MZI based on a SMF and two peanut taper joints. As the figure shows, the guided light is split into two beams propagating in the cladding and core respectively at the first peanut joint. The SMF in the middle serves as a functional fiber where the light in the cladding is easily affected by the surrounding environment. The interference happens when the two beams meet at the second joint. Because of the RI difference between the core and cladding, there is a phase difference between the two beams. The phase difference can be expressed as

$$\Delta\varphi = \frac{2\pi \, \Delta n_{\text{eff}} \, L_{\text{eff}}}{\lambda} \tag{1}$$

where L_{eff} is the effective length of MZI, which is usually the geometry length of the functional fiber, λ is the light wavelength, Δn_{eff} is the effective RI difference between the core mode and the cladding mode. $\Delta n_{eff} = (n_{co} - n_{cl})$, where n_{co} and n_{cl} are the effective RI of the core mode and the cladding mode, respectivley. The intensity of light at the lead-out fiber is

$$I_{out} = I_{co} + I_{cl} + 2\sqrt{I_{co} I_{cl}} \cos(\Delta\varphi) \tag{2}$$

where I_{co} and I_{cl} are the light intensity in the core and cladding. Figure 1b shows the transmission spectrum of the SMF-based MZI with a 5 cm SMF. The dips in the transmission spectrum appear where the phase-match condition ($\Delta\varphi = (2k + 1))\pi$,

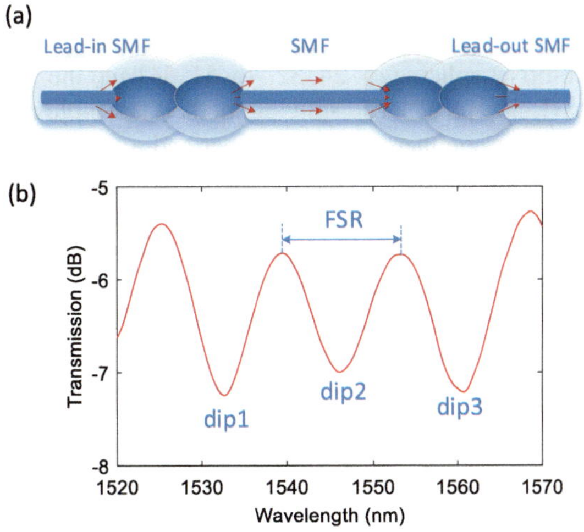

Fig. 1 **a** Schematic diagram and **b** transmission spectrum of an in-fiber MZI based on an SMF and two peanut joints

k is an integer) is meet. Therefore, the wavelengths of the dips are determined by

$$\lambda_{dip} = \frac{\Delta n_{eff} L_{eff}}{2k + 1} \tag{3}$$

The free spectrum range (FSR) [28] and the fringe visibility (K) [27] are given as

$$FSR \approx \frac{\lambda^2}{\Delta n_{eff} L_{eff}} \tag{4}$$

$$K = \frac{2\sqrt{I_{co} I_{cl}}}{I_{co} + I_{cl}} \tag{5}$$

The optical path difference (OPD, i.e. $\Delta n_{eff} L_{eff}$) between the core mode and the cladding mode is easily affected by the surrounding medium. Specifically, the RI of the surrounding medium can directly affect n_{cl}. Temperature and strain applied to the functional fiber can change n_{cl} due to the thermo-optical effect and the elastic-optic effect, respectively. Curvature applied on the fiber will lead to a change of L_{eff}. Therefore, the variation of these measurands can directly induce a change to the output intensity and the wavelength shift of spectrum dips. In other words, in-fiber MZIs can detect these measurands by monitoring the change of I_{out} and λ_{dip}. As for other measurands, such as humidity, magnetic field and gas or ion concentration, they can be detected by the in-fiber MZI sensors as long as materials are incorporated to transform these measurands to the RI or temperature exerted on the functional fiber.

2.2 *Fabrication Methods*

There are mainly four categories of fabrication methods for in-fiber MZIs. The first type is the splicing method, including the core-mismatch splicing [2, 7, 41, 47], the core-offset splicing [5, 24, 32, 46, 51, 54] and the joint-deformation splicing [8, 22, 30, 31, 35, 42, 52] The splicing method is simple and cost-efficient because it only involves a commercial fiber cleaver and a fusion splicer. It worth mentioning that, the fusion parameters, such as the fusion current, fusion time and the gap between the cleaved fibers need to be modified when it comes to different types of fibers and splicing techniques [42]. Figure 2 exhibits several in-fiber MZIs based on the three splicing techniques mentioned ahead. Figure 2a, b show two in-fiber MZIs made by the core-mismatch splicing technique. As they present, the functional fiber and the two fibers on its both sides have different core diameters so that the input light can be split into two beams to allow interference to happen. In most cases, the two beams travel in the core and cladding of the functional fiber separately [8, 22, 30, 35]. In some cases, for instance, when the functional fiber is a dual-core fiber (DCF), the input light can be split into the two cores of the DCF so that the interference happens between the two core modes [2]. Figure 2 **c** shows the in-fiber MZIs based on the core-offset

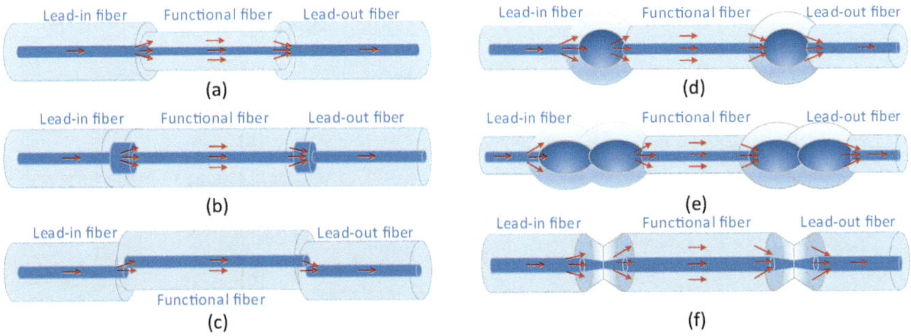

Fig. 2 In-fiber MZIs based on **a, b** core-mismatch splicing, **c** core-offset splicing and **d–f** joint-deformation splicing

splicing technique, which shows a lateral offset between the functional fiber and the lead-in/lead-out fiber. Figure 2d–f present the in-fiber MZIs based on three different deformed joints, including the spherical joint [30, 52], peanut joint [8, 42] and the tapered joint [35]. As with simple splicing technique, the purpose of introducing a core-offset to the splicing point or splicing the joint into deformed shapes is also to distribute the input light into the core and the cladding of the functional fiber. Compared with the simple splicing technique, the latter two splicing techniques are more flexible since they can construct MZIs with fibers that have both matched and mismatched cores.

The second type is the femtosecond (fs) laser inscribing method where fs lasers are applied to inscribe microchannels in the functional fibers to create dual paths for the guided light to travel [6, 48, 49, 50]. Figure 3 shows three in-fiber MZIs made by the fs laser inscribing method. In Fig. 3a, the in-fiber MZI is a SMF-NCF (no core fiber)-SMF structure where a straight refractive index modified waveguide is inscribed in the NCF by the fs laser [48]. In Fig. 3a, b bridge-shaped refractive index modified waveguide is inscribed in a SMF to form a two-beam interference channel together with the fiber core of the SMF [49, 50]. Figure 3c shows an in-fiber MZI based on a SMF-GIF (graded index fiber)-SMF structure where an air cavity is ablated in the GIF by the fs laser [6].

The third type is the microtapering method which involves the microtapering of the functional fiber. Figure 4 shows an in-fiber MZI based on a microtapered SMF. The distinct property of the microfiber MZI is that part of the light will propagate into the surrounding medium along the microtapered region. Therefore, the microfiber MZI shows very high sensitivity to the surrounding medium [29].

The last type is the macrobending method, which utilizes macrobends as the light splitting/combing structure. Figure 5 shows two in-fiber MZIs based on SMF with macrobends [1, 28]. Figure 5a shows a θ shaped structure and Fig. 5b shows a droplet bend structure. The splitting and combining of the light beam happen at the macrobend area. The fabrication of such structure is quite simple and cost-effective since the macrobends needs no extra device and can be made manually. It worth mentioning that the coating of the SMFs used in the two MZIs are unstripped to

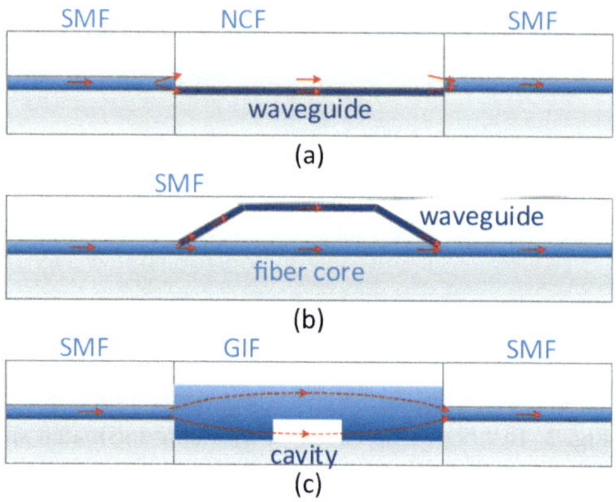

Fig. 3 **a–c** microchannel-based in-fiber MZIs inscribed using fs laser [6, 48, 49, 50]

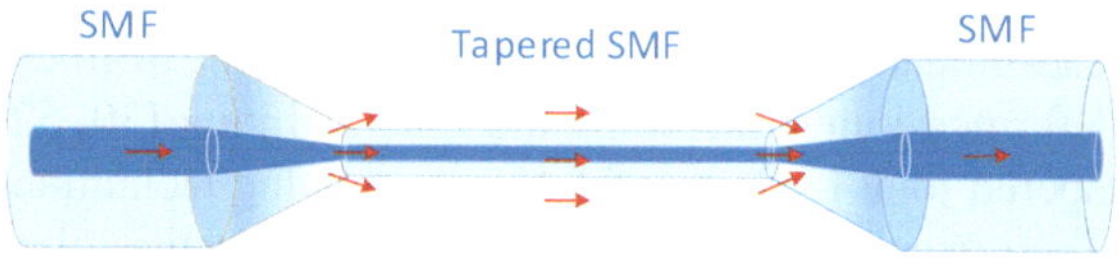

Fig. 4 An in-fiber MZI with tapered SMF

protect the microbends. Therefore, the sensing performance will be influenced by the coating material.

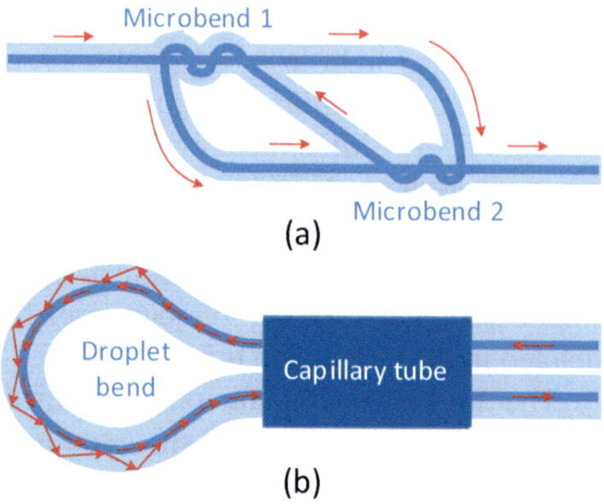

Fig. 5 **a, b** In-fiber MZIs based on SMF with macrobends [1, 28]

3 Experimental Study in Sensing Applications/Sensing Applications

Thanks to the versatile fabrication methods and flexible structures, in-fiber MZIs can be utilized for the sensing of various measurands, such as temperature, strain, curvature, RI, humidity and the concentration of gas and metal ions. In this section, recent progress in these sensing applications of the in-fiber MZIs is reviewed.

3.1 In-Fiber MZIs for Temperature Sensing

Temperature measurement plays an important role in various industries since temperature is a crucial indicator in industrial process monitoring. For instance, in food industries, temperature measurement is crucial to control bacterial growth and maintain the quality of the processed products. In biochemistry, temperature monitoring is vital in the production, transportation and storage of vaccines, medicine, detection reagent and biological samples. At present, most temperature sensors on the market are electric sensors. Fiber-optic sensors, including in-fiber MZI sensors, provide a new approach for temperature sensing. Compared with their electric counterparts, fiber-optic temperature sensors have the advantages of compact size, anti-electromagnetic interference, anti-radiation and the capability of multiplexing. They have great potential in harsh environments such as places with strong radiation or electromagnetic interference.

When temperature changes in the surroundings, the refractive index and the effective length of the in-fiber MZI will be affected because of the thermo-optic effect and the photoelastic effect. Take the SMF-TCF-SMF structure as an example, the wavelength of the spectrum dip will shift as the temperature changes and the sensitivity of dip wavelength to temperature is [44]

$$\frac{\Delta\lambda_{dip}}{\Delta T} \approx \lambda_{dip}(\alpha + \beta) \tag{6}$$

where α and β are the effective thermo-optic coefficient (TOC) and the thermo-expansion coefficient (TEC) of the fiber.

According to Eq. 6, the temperature sensitivity of the in-fiber MZI depends on the material of the fiber. For common silica fibers, the sensitivity of the dip wavelength depends mostly on TOC of the silica and is around dozens of pm/°C. For instance, recently, two in-fiber MZIs based on an SMF-NCF-SMF structure and an SMF-TPMF (thin polarization maintaining fiber) -SMF structure are proposed [34, 44]. Their sensitivities are 51.3 pm/°C and 91.84 pm/°C, respectively.

The sensitivity can be enhanced by raising the effective TOC of the functional fiber. This can be done by two categories of methods. The first one is to utilize functional fiber with higher TOC. For instance, highly doped silica fibers tend to have higher

TOC than the common germanium (Ge) doped silica fibers. In 2019, R. H. Fan proposed an in-fiber MZI based on a boron (B) and Ge co-doped photosensitive fiber for temperature sensing [8]. It has a temperature sensitivity of −386.96 pm/°C in the range of 45–65 °C. Recently, an in-fiber MZI based on an erbium (Er)-doped fiber (EDF) is proposed and utilized in a fiber ring laser [22]. Its temperature sensitivity is 301 pm/°C in the range of 0–50 °C.

The other more commonly utilized method is applying materials with higher TOC to the functional fiber. These materials will add to the overall effective TOC of the in-fiber MZI sensor. There are several approaches to apply the materials to the functional fiber. The first one is to coat the functional fiber with such materials. The purpose of this approach is to use the temperature-sensitive material to affect the effective RI of the cladding mode in the functional fiber. For instance, a microfiber MZI coated with PDMS is proposed for temperature sensing [36].PDMS is a temperature-sensitive organosilicon polymer material which is adhesive with silica materials. The TOC of PDMS is -4.69×10^{-4} RIU/°C, two orders higher than the silica fibers. In the range of 20–100 °C, the PDMS-coated microfiber MZI has a sensitivity of 220 pm/°C, 5.2 times higher than the uncoated structure. The diameter of the microtaper region is 58 μm in the demonstrated experiment. The sensitivity can be further enhanced by tapering the functional fiber thinner. In 2021, S. Cheng et al. proposed an in-fiber MZI based on a tapered multicore fiber (TMCF) [4]. The TMCF is also coated with PDMS and has a waist diameter of 7 μm. Its temperature sensitivity reaches up to 25 nm/°C.

The second approach is to utilize fiber with holes, such as hollow core fiber (HCF), as the functional fiber and fill the holes with temperature-sensitive material. In 2018, a polymer structure is printed in a HCF by fs laser to construct an in-fiber MZI for high-sensitivity temperature sensing. The sensitivity reaches up to 6.5 nm/°C in the range of 25–30 °C [17]. Apart from polymer materials, liquids with high TOC are also used. In 2021, an in-fiber MZI based on an HCF filled with liquid crystal is proposed with a temperature sensitivity of 1.318 nm/°C [21]. Recently, another in-fiber MZI based on an HCF filled with Cargille-liquid is proposed. The TOC of the liquid is -4.07×10^{-4} RIU/°C, which enables the in-fiber MZI sensor a sensitivity of 13.87 nm/°C in the range of 25–50 °C [16]. The sensor structure, experimental setup as well as the spectrum response and dip wavelength shift with different temperatures are shown in Fig. 6. Besides HCF, other fibers with holes are also used. Recently, a DCF with a center core, an eccentric core and a hole near the center core is proposed to construct a SMF-MMF-DCF-MMF-SMF structure for MZI-based temperature sensing [26]. The hole near the center core is filled with glycerol-water solution which has a TOC of -2.2×10^{-4} RIU/°C and enables the in-fiber MZI to have a sensitivity of 5.69 nm/°C in the range of 25–40 °C.

Another approach is to fabricate a channel or cavity in the functional fiber and then fill the channel or cavity with temperature-sensitive materials [12, 43, 45]. In 2018, H. Zhang et al., proposed an in-fiber MZI sensor based on two air cavities fabricated from two sections of D-shaped fiber. The whole sensor is immersed in RI liquid for temperature sensing. The TOC of the RI liquid is 3.95×10^{-4} RIU/°C. It is found that the sensitivity of the MZI sensor reaches up to −84.72 nm/°C when RI of the

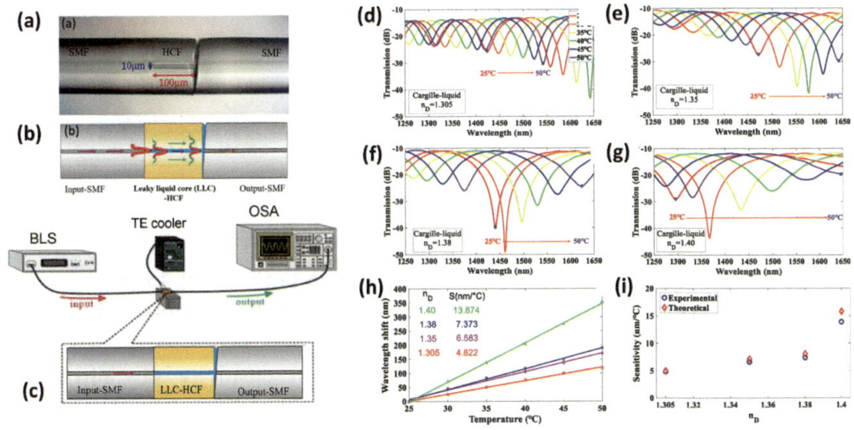

Fig. 6 **a**, **b** Schematic diagram of the liquid-filled HCF-based in-fiber MZI sensor structure; **c** experimental setup of the sensor; **d–i** spectrum response and dip wavelength shift with different temperature [16]

liquid is 1.482 [45]. In 2019, F. Zhang et al. proposed an SMF-GIF-SMF structure with an air cavity inscribed in the GIF and filled with UV-curable polymer. This polymer-filled in-fiber MZI structure has a temperature sensitivity of −13.27 nm/°C in the range of 22–42 °C [43].

3.2 *In-Fiber MZIs for the Sensing of Strain and Curvature*

Strain sensors are utilized to check the strength and health of buildings, civil structures and industrial products. Many in-fiber MZIs are proposed and demonstrated for strain sensing. When axial strain applies on the in-fiber MZI sensor, the fiber structure will be elongated and the effective RI of the core and cladding modes will change because of the photoelastic effect. Therefore, the wavelength of the dips in the transmission spectrum of the in-fiber MZI sensor change accordingly. Take the in-fiber MZI based on a SMF and two peanut joints in Fig. 1 as an example again, the sensitivity of its dip wavelength is determined by [37]

$$\frac{\Delta\lambda_{dip}}{\Delta\varepsilon} = \lambda_{\text{dip}}\left(\frac{1}{L_{eff}}\frac{\partial L_{eff}}{\partial\varepsilon} + \frac{1}{\Delta n_{eff}}\frac{\partial\Delta n_{eff}}{\partial\varepsilon}\right) \tag{7}$$

where $\Delta\varepsilon = \frac{\Delta L_{eff}}{L_{eff}}$ is the variation of axial strain applied on the fiber. $\frac{\partial L_{eff}}{\partial\varepsilon}$ is the strain-induced effective length variation whereas $\frac{\partial\Delta n_{eff}}{\partial\varepsilon}$ is the strain-induced RI difference variation. For most silica-fiber based in-fiber MZIs, the strain sensitivity is around several pm/με. For instance, in 2019, X. Liang et al. proposed an in-fiber MZI sensor based on core-mismatch splicing technique is proposed. The sensor utilizes an SMF

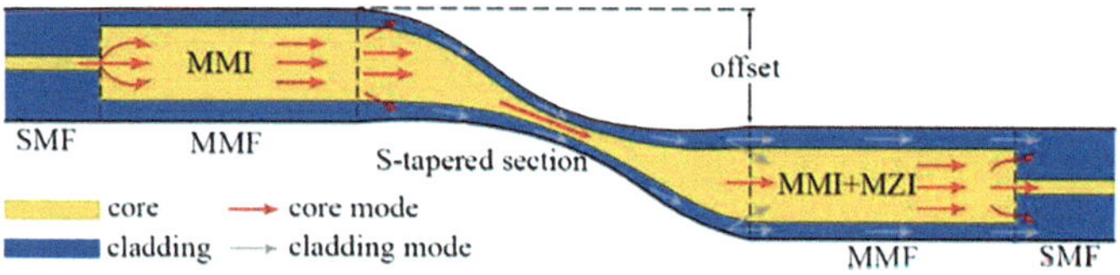

Fig. 7 Schematic diagram of a strain sensor based on a SMF-tapered MMF-SMF structure [33]

as the functional fiber and its strain sensitivity is 1.78 pm/με [20]. In 2020, another in-fiber MZI sensor based on core-offset splicing and a PCF is proposed with a strain sensitivity of 1.97 pm/με [39].

The most commonly used way to enhance the strain sensitivity is by introducing structures with smaller diameters, such as tapered or ablated structures, to the functional fiber. The fiber region with a smaller diameter has a smaller cross-sectional area. In this case, the strain applied on the whole sensor is more distributed to this thinner region. Therefore, both $\frac{\partial \Delta n_{eff}}{\partial \varepsilon}$ and $\frac{\partial L_{eff}}{\partial \varepsilon}$ are larger in the thinner region, which enables a higher strain sensitivity [38]. For instance, in 2018, a SMF-tapered MMF-SMF structure is proposed for strain sensing [33]. Figure 7 shows the structure of the MZI sensor. As the figure shows, the middle part of the MMF is tapered to an S-shape to enable a higher sensitivity. Because of the S-tapered structure, the in-fiber MZI sensor has a sensitivity of −103.8 pm/με, 33 times better than the untapered structure. In 2021, a bent SMF fabricated by laser ablating is utilized for strain sensing [38]. A section of the SMF is partially ablated to have a thinner diameter and then part of the ablated region is bent to obtain MZI. The sensor has a rather high sensitivity of 165 pm/με in the region of 0–200 με. In 2022, an in-fiber MZI sensor based on a SMF and two core-offset microtapered joints is proposed. The two microtapered joints allow the sensor to have a higher sensitivity. It has a strain sensitivity of – 108.0 pm/με in the range of 0–400 με and a sensitivity of – 47.6 pm/με in the range of 400–1200 με [37].

In-fiber MZIs are also proposed for the sensing of curvature or bending deformation. Bending deformation is an important measurand in mechanical engineering, robot control, aerospace applications, structural health monitoring and even human health monitoring. For instance, curvature sensors can be used to monitor the gesture motion of robot arms. They can also be used to monitor human posture motion for medical applications. For instance, the measurement of knee-joint bending can help assess some crucial health issues such as rehabilitation from sports injuries and some chronic diseases related to brain degeneration [5].

Figure 8 shows an experimental setup for curvature sensing. As the figures shows, the in-fiber MZI sensor is placed on two translation stages. The curvature applied on the sensor is adjusted by modulating the distance between the two stages. It can be approximately expressed by [15, 31]

$$C = \frac{1}{R} \approx \sqrt{\frac{24x}{L_0^3}} \tag{8}$$

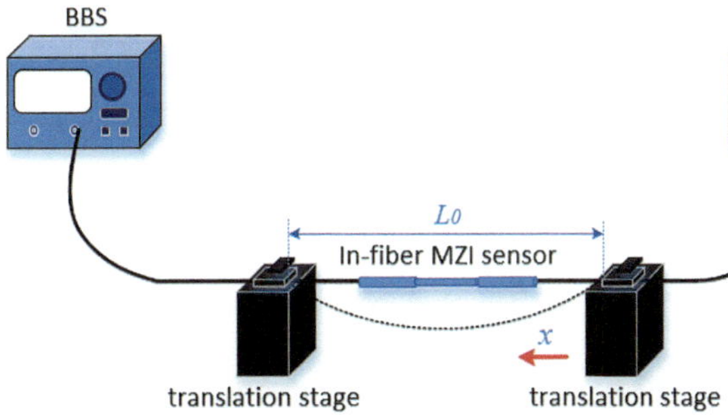

Fig. 8 Experimental setup for curvature sensing (BBS: broadband source, OSA: optical spectrum analyzer)

where R is the bending radius, x is the displacement distance of the translation stage, L_0 is the length of the original distance between the two stages.

When the sensor is bent, a strain $\varepsilon = d \cdot C$ will be applied on the sensor, where d is the distance between the core mode and the cladding modes. As curvature C changes, the bent-induced strain varies. The strain variation $\Delta\varepsilon$ will induce a variation to the effective RI difference between the core mode and the cladding modes $\left(\Delta n_{eff}\right)$. The effective RI difference variation can be expressed as $\partial \Delta n_{eff} = k\Delta\varepsilon = kd\Delta C$, where k is the strain-RI coefficient. Therefore, the wavelength shift of the spectrum dips induced by the variation of the curvature is approximately [15, 31]

$$\Delta\lambda \approx \frac{\lambda_{dip}}{\Delta n_{eff}} kd\Delta C \tag{9}$$

For many in-fiber MZI sensors proposed for curvature sensing, their curvature sensitivities ranges from several to dozens of nm/m^{-1}. For example, recently, three in-fiber MZIs based on laser-inscribed RI modulated channels are proposed for curvature sensing [27, 47, 48]. Among the three in-fiber MZI sensors, the SMF-based and the FMF-based laser-inscribed sensor separately have a sensitivity of 2.491 nm/m^{-1} and 2 nm/m^{-1} in the range of 0–10 m^{-1}. The NCF-based laser-inscribed sensor has a sensitivity of 14.91 nm/m^{-1} in the range of 0–3 m^{-1}.

As with strain sensing, the sensitivity of the in-fiber MZI proposed for curvature sensing can be enhanced by decreasing the diameter of the functional fiber. For instance, in 2021, an in-fiber MZI based on a tapered SCF (seven core fiber) achieved a high sensitivity of 174.02957 nm/m^{-1} in the range of 0–0.04702 m^{-1} [14]. Recently, X. Zhu et al. proposed a tapered NCF-based in-fiber MZI for curvature sensing. Figure 9 shows the schematic diagram and the micrograph of the curvature sensor. As the figure shows, the central waist region of the sensor is tapered to a rather thin diameter. With the tapered sensing region, the sensor has a high sensitivity of −224.64 nm/m^{-1} in the range of 0–0.052 m^{-1} [53].

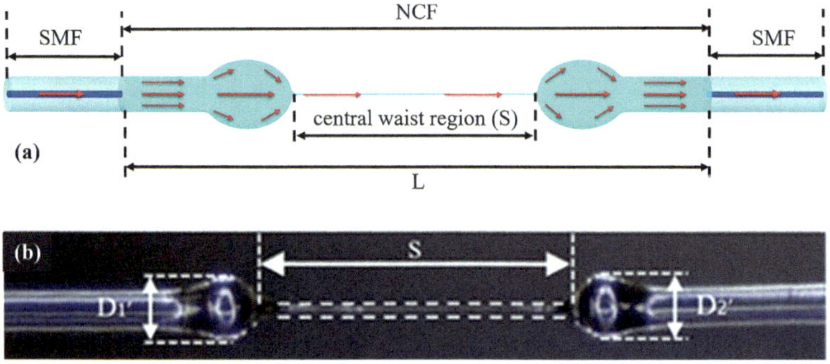

Fig. 9 **a** Schematic diagram and **b** micrograph of the in-fiber MZI sensor based on a tapered NCF [53]

3.3 *In-Fiber MZIs for RI Sensing*

Refractive index sensors, or refractometers, play an important role in biological, chemical and medical industries. By measuring the RI of a liquid sample, we can recognize its identity, assess its purity or determine its concentration. In recent years, numerous in-fiber MZIs are proposed for RI sensing [3, 7, 40]. For the in-fiber MZI sensors, when RI in the surrounding changes, a variation will be caused to the effective RI of the cladding mode and even that of the core mode. Therefore, the effective RI difference between the core mode and the cladding mode (Δn_{eff}) varies with the ambient environment as well. The sensitivity to surrounding RI (n_{sr}) of the spectrum dip can be expressed as [40]

$$\frac{\Delta\lambda_{dip}}{\Delta n_{sr}} = \frac{\lambda_{dip}}{\Delta n_{eff}} \frac{\partial \Delta n_{eff}}{\partial n_{sr}} \tag{10}$$

For many in-fiber MZIs, the diameter of their functional fibers is approximately the diameter of the standard SMF. The RI variations of the surrounding have limited influence on the effective RI of the cladding mode which is still much larger than that of the core mode. For these sensors, their RI sensitivities are in the level of dozens of nm/RIU (RI unit) to several hundreds of nm/RIU. For instance, in 2019, an in-fiber MZI based on a PCF with an outer diameter of 125 μm is proposed for RI sensing [7]. It has a sensitivity of 93.33 nm/RIU in the RI range of 1.333–1.381 with a functional fiber length of 35 mm. By further etching the PCF to a thinner diameter of 91 μm, its sensitivity reaches 359.37 nm/RIU. Except from etching, other techniques to enhance the RI sensitivity of the in-fiber MZI sensors include using microfibers or fibers with open cavity that can allow the surrounding fluid to flow in. The principle of these techniques is to allow the RI of surrounding medium to affect effective RI of the cladding mode or the core mode as much as possible. For instance, in 2020, F. Xia et al. proposed a microfiber MZI with a waist diameter of 4.8 μm for RI sensing [40]. With such a thin diameter, the surrounding medium can easily affect the cladding mode and the core mode. Figure 10 shows the schematic diagram and the micrographs of the microfiber MZI sensor. As the figure shows, there are two dominant modes excited in the sensing area, including the HE_{11} mode and the HE_{12} mode. The sensor has an ultra-high RI sensitivity of 24,209 nm/RIU in the range of 1.3320–1.3345. Except for RI sensing, it can also be used for temperature sensing and has a higher sensitivity when it is coated with PDMS. Recently, a microcavity with a depth of 62.5 μm and a width of 180 μm is fabricated by a fs laser in a SMF to constitute an in-fiber MZI RI sensor. The microcavity allows the fluid around the sensor to access the core of the functional region. The RI sensitivity of the sensor is up to 16,660 nm/RIU in the range of 1.3633–1.3783 [3].

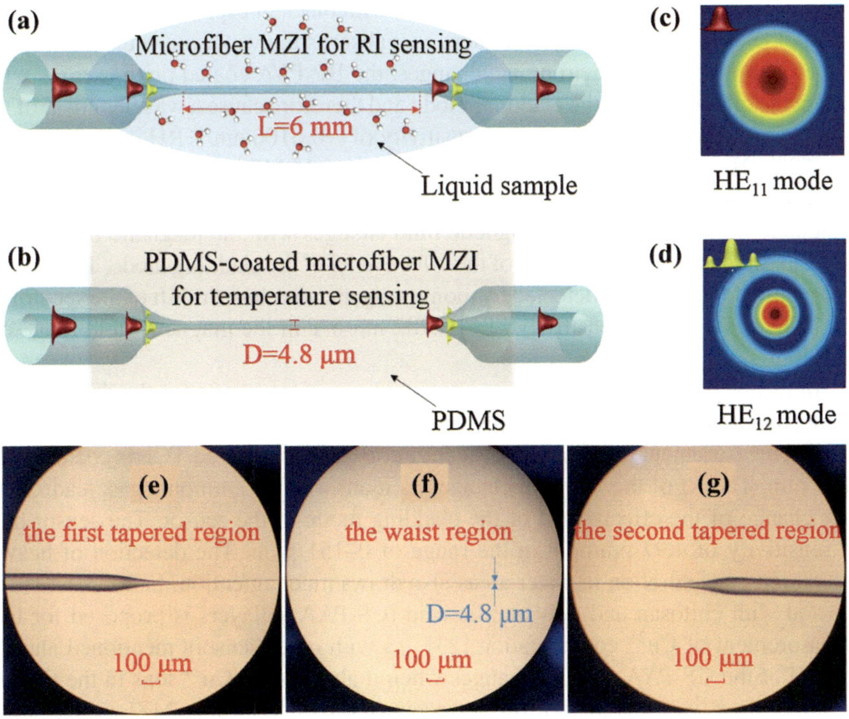

Fig. 10 Schematic diagram of the microfiber MZI **a** for RI sensing and **b** for temperature sensing, **c** H_{11} mode and **d** H_{12} mode excited in the sensor; Micrographs of **e** the first tapered region, **f** the waist region and **g** the second tapered region of the microfiber MZI. [40]

3.4 In-Fiber MZIs for the Sensing of Some Other Measurands

The in-fiber MZIs are also proposed for the sensing of humidity, magnetic field and the concentration of gas and metal ion [6, 9, 10, 23, 30]. Generally speaking, the in-fiber MZIs themselves are not sensitive to these quantities. In order to make them sensitive, materials that can react with these quantities are coated onto or filled into the functional fiber to transform the measurands to the RI or temperature variations applied on the functional fiber.

Materials whose RI changes when react with the measurands are most commonly used. Recently, an in-fiber MZI based on a TCF and two spherical joints is proposed for humidity sensing [30]. The TCF is coated with black phosphorus (BP) -polyvinyl alcohol (PVA). RI of the BP-PVA alters as it absorbs the water molecules in the air. When humidity in the environment changes, the BP-PVA coated on the sensor will affect the effective RI of the cladding mode and therefore cause a wavelength shift to the spectrum dip. The sensor has a sensitivity of −0.6106 nm/%RH in the range of 40–50%RH. Another in-fiber MZI based on a tapered SMF is proposed for magnetic field sensing [10]. The tapered SMF is sealed in a capillary tube which is filled with magnetic fluid. The RI of the magnetic fluid changes with the magnetic field in the surrounding and lead to a change of the effective RI of the cladding mode. Therefore, the magnetic field can be detected by monitoring the wavelength shift of the spectrum dip. The sensor has a sensitivity of − 5.782 nm/mT in the magnetic field range of 10–30 mT.

In recent years, quite a few fiber-optic sensors are proposed for the detection of toxic, flammable and explosive gases. In 2021, a tapered TCF-based MZI is proposed for the measurement of ammonia concentration [9]. The tapered TCF is coated with GO film. The RI of the GO film changes as it absorbs the ammonia gas, leading to the change of the effective RI of the cladding mode in the sensor. The sensor has a sensitivity of 4.97 pm/ppm in the range of 0–151 ppm. The detection of heavy metal ion concentration in water areas also draws much attention. In 2022, an HCF coated with chitosan and poly acrylic acid (CS-PAA) bilayers is proposed for the measurement of Cu^{2+} concentration [23]. As with other sensors mentioned ahead, the RI of the CS-PAA coating changes when it absorbs the Cu^{2+} ions in the liquid, resulting in a wavelength shift of the spectrum dip. The in-fiber MZI sensor has a sensitivity of 42 nm/mM in the range of 0–40 μM.

Materials that can alter the temperature around the functional fiber of the in-fiber MZI are also proposed and demonstrated. In 2021, B. Du et al. proposed a SMF-GIF-SMF structure for the measurement of hydrogen concentration in the air [6]. Figure 11a shows the structure of the hydrogen sensor. As the figure shows, an air cavity is ablated in the GIF by fs laser and filled with PDMS. Then the sensor is coated with Pt-loaded tungsten trioxide (WO_3). When hydrogen meets the Pt-loaded WO_3 on the surface of the in-fiber MZI sensor, the exothermic reaction occurs and leads to a temperature rise, resulting in a RI change of the filled PDMS. Figure 11b, c show the transmission spectrum evolution with the increase of hydrogen concentration and

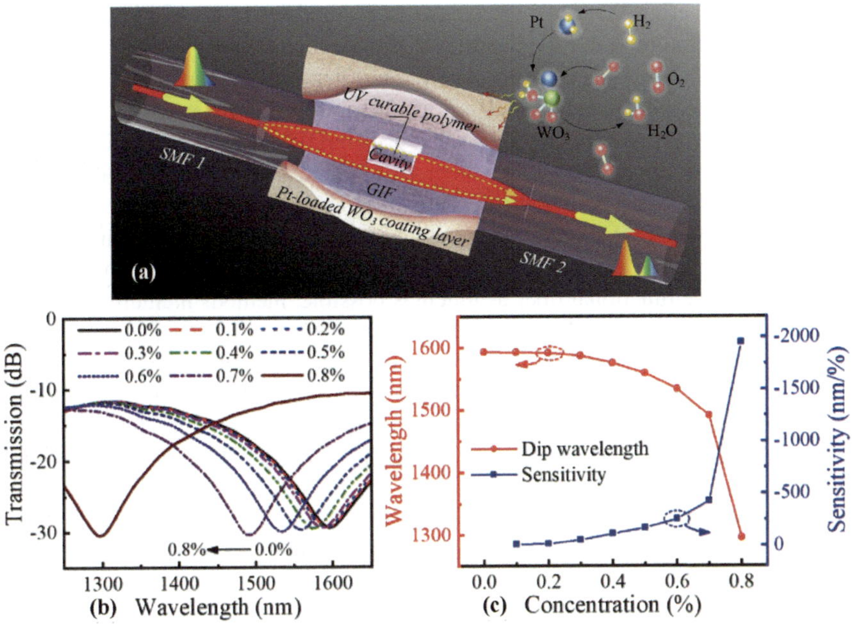

Fig. 11 **a** Schematic diagram and sensing principle of the in-fiber MZI hydrogen sensor, **b** the transmission spectrum evolution with different hydrogen concentrations and **c** spectrum dip wavelength and sensitivity as funciton of hydrogen concentrations [6]

the dip wavelength and sensitivity as a function of the hydrogen concentration. As Fig. 11c shows, the sensor has a maximum sensitivity up to −1948.68 nm/% (vol %) in the hydrogen concentration range of 0%–0.8%.

4 Conclusion

In this chapter, basic principle, fabrication methods, sensing mechanisms and sensing applications of the in-fiber MZIs are summarized. To conclude, the working principle of the in-fiber MZIs is based on the interference between two light beams inside the functional fiber, which for most cases are the beam of the core mode and the beam of the cladding mode. There are mainly four methods that are most commonly used to fabricate the in-fiber MZIs, including the splicing method, the fs-laser inscribing method, the microtapering method and the macrobending method. With these methods, a large number of in-fiber MZIs are proposed and demonstrated for the sensing of various measurands. Temperature, strain, curvature and RI can exert direct influence to the wavelength shift of the transmission spectrum dips of the in-fiber MZI sensors by changing the OPD of the sensor. Other measurands such as humidity, magnetic field, harmful gases and metal ions can exert indirect influences by applying materials that react with these measurands and then add to a change to the RI or temperature in the surrounding of the functional fiber.

Methods to enhance the sensitives of the in-fiber MZI sensors are also summarized. For temperature sensing applications, the most effective way to enhance the sensitivity is to apply temperature-sensitive materials to the functional fiber. For the sensing of strain and curvature, methods to decrease the diameter of the functional fiber, including tapering, etching and laser ablating are most commonly used to enhance the sensitivities. As to the sensing of the liquid RI, the key to improve the sensitivity is to allow the surrounding liquid to influence the cladding mode or the core mode of the functional fiber as much as possible. methods including tapering, etching and using fibers with open cavities are applied. For the sensing of humidity, magnetic field, gas concentration and metal ion concentration, the sensitivities rely much on the materials that are used. Most of these materials affect the interference by altering the effective RI of the cladding mode or the core mode. By combining proper functional materials with tapering, etching and laser ablating, the sensitivity of these sensors can be greatly enhanced.

Thanks to different fabrication methods as well as numerous functional fibers and materials, in-fiber MZIs are quite versatile and suitable for various sensing applications. Apart from the sensing applications summarized in this chapter, attempts of biosensing and acoustic sensing with the in-fiber MZIs are also witnessed [11, 18, 25]. Still, more sensing applications of the in-fiber MZIs are to be discovered if appropriate structures and materials are chosen. Moreover, beside the enhancement of sensitivity, more efforts should be made to improve other sensing properties such as sensing accuracy, repeatability and measurement range. For instance, as with other fiber-optic sensors, in-fiber MZI sensors suffer the vulnerability to multiple physical quantities in the environment, which hinders the accuracy and repeatability. How to exclude other physical quantities except for the quantity to be measured remains an open issue. In addition, there is a trade-off between the sensitivity and the measurement range of the in-fiber MZI sensors. To solve this problem, a segmentation method which utilizes an FBG to determine the measurement range sections is proposed to enlarge the measurement range without hindering the sensitivity [19]. Another signal demodulation method which analyzes the signal in the frequency-domain can also help solve this problem [13]. However, new methods and schemes are to be found to provide alternative and better solutions to this problem. In conclusion, there is still much work to be done for the development of in-fiber MZIs for sensing applications.

References

1. Al-Hayali SK, Salman AM, Hadi Al-Janabi A. Effect of hygroscopic polymer-coatings on the performance of relative humidity sensor based on macro-bend single-mode fiber. Opt Fiber Technol. 2021;62: 102460.
2. Chen H, Hu X, Chen X, Yu Q, Lian Z, Wang H, Qu H. In-Line interferometric temperature sensor based on dual-core fiber. IEEE Sens J. 2021;21:12146–52.

3. Chen N, Liu C, Lu Z, Tao W, Peng M. Femtosecond laser processing for a high sensitivity fiber MZI microcavity. Opt Express. 2022;30:12397–408.
4. Cheng S, Hu W, Ye H, Wu L, Li Q, Zhou A, Yang M, Zhao Q, Guo D. Tapered multicore fiber interferometer for ultra-sensitive temperature sensing with thermo-optical materials. Opt Express. 2021;29:35765–75.
5. Ding L, Yu L, Hu G, Xie Q, Zhu L. Knee joint curvature detection system based on fiber optic mach-zendler interferometric curvature sensor. IEEE Sens J. 2021;21:28017–24.
6. Du B, He J, Yang M, Wang Y, Xu X, Wang J, Zhang Z, Zhang F, Guo K, Wang Y. Highly sensitive hydrogen sensor based on an in-fiber Mach-Zehnder interferometer with polymer infiltration and Pt-loaded WO(3) coating. Opt Express. 2021;29:4147–58.
7. Du H, Sun X, Hu Y, Dong X, Zhou J. High sensitive refractive index sensor based on cladding etched photonic crystal fiber mach-zehnder interferometer. Photonic Sensors. 2019;9:126–34.
8. Fan RH, Li LQ, Wang CC, Deng YD, Lv XX, Ren ZJ, Shen JG, Peng BJ. Interference-type high-sensitivity temperature sensor based on the photosensitive fiber microsphere. IEEE Sens J. 2019;19:11941–5.
9. Fan X, Deng S, Wei Z, Wang F, Tan C, Meng H. Ammonia gas sensor based on graphene oxide-coated mach-zehnder interferometer with hybrid fiber structure. Sensors (Basel). 2021;21.
10. Gu S, Feng D, Zhang T, Deng S, Li M, Hu Y, Sun W, Deng M. Highly sensitive magnetic field measurement with taper-based in-line mach-zehnder interferometer and vernier effect. J Lightwave Technol. 2022;40:909–17.
11. Han X, Li C, Hu Y, Li J, Gao F, Zhang W, Dong X, Fang B, Zhang G, Xu J. Compact dynamic in-fiber acoustically-induced mach-zehnder interferometer based on phase mismatch and its application in a tunable and switchable dual-wavelength laser. J Lightwave Technol. 2021;39:3539–45.
12. He J, Zhang F, Xu X, Du B, Wu J, Li Z, Bai Z, Guo J, Wang Y, He J. Highly sensitive temperature sensor based on cascaded polymer-infiltrated fiber mach-zehnder interferometers operating near the dispersion turning point. Polymers (Basel). 2022;14.
13. Hu PB, Chen ZM, Yang M, Yang JY, Zhong C. Highly sensitive liquid-sealed multimode fiber interferometric temperature sensor. Sensor Actuat a-Phys. 2015;223:114–8.
14. Ji Y, Sun D, Chen Y, Shi Y, Cao J, Zhang G, Han Z, Wang C, Zhu X. A high sensitivity curvature sensor based on microfiber mach-zehnder interferometer with tapered seven-core fiber. IEEE Sens J. 2021;21:24090–7.
15. Jiang B, Bai Z, Wang C, Zhao Y, Zhao J, Zhang L, Zhou K. In-Line mach-zehnder interferometer with d-shaped fiber grating for temperature-discriminated directional curvature measurement. J Lightwave Technol. 2018;36:742–7.
16. Lee CL, Zhuo WR, Liu TK. Highly modulated in-fiber mach-zehnder interferometer based on an ultracompact leaky-guided liquid core. Sensors (Basel). 2022;22.
17. Li C, Liao C, Wang J, Gan Z, Wang Y. Femtosecond laser microprinting of a polymer optical fiber interferometer for high-sensitivity temperature measurement. Polymers (Basel). 2018;10.
18. Li H-C, Leng Y-K, Liao Y-C, Liu B, Luo W, Liu J, Shi J-L, Yuan J, Xu H-Y, Xiong Y-H, et al. Tapered microfiber MZI biosensor for highly sensitive detection of staphylococcus aureus. IEEE Sens J. 2022;22:5531–9.
19. Li M, Gong Y, Yin J, Li W, Shao Y, Cong A, Huang G. Highly-sensitive and wide-range temperature sensor based on polymer-filled micro-cavity in fibre Bragg grating by temperature segmentation. Optik. 2021;245: 167707.
20. Liang X, Ning T, Li J, Li Y, Liu Z. Axial micro-strain sensor based on resonance demodulation technology via dual-mode CMECF. Photonic Sensors. 2018;9:78–88.
21. Lin W, Zhou S, Liu Y, Vai MI, Shao L. Liquid crystal-embedded hollow core fiber temperature sensor in fiber ring laser. Appl Sci. 2021;11:7103.
22. Lin W, Zhou S, Shao L-Y, Vai MI, Shum PP, Zhao F, Liu S, Hu J, Liu Y. In-fiber mach-zehnder interferometer based on er doped up-taper and peanut-shaped fiber structure in fiber ring laser. IEEE Access. 2021;9:128126–32.
23. Liu MF, Wang JW, Hwang SJ. In-fiber Mach-Zehnder interferometer based on hollow optic fiber for metal ion detection. Opt Express. 2022;30:26006–17.

24. Mu X, Yang Y, Wang J, Bi L. High sensitive Mach-Zehnder interferometer for seawater refractive index measurement. Ferroelectrics. 2022;595:27–34.
25. Ojha NNS, Kumar N. A simple technique for detecting the broad range of organic/inorganic materials. Opt Fiber Technol. 2021;64: 102542.
26. Qiu H, Zhao C, Hu X, Chen H, Yu Q, Lian Z, Qu H. Glycerol–water solution-assisted mach-zehnder temperature sensor in specialty fiber with two cores and one channel. Photonics. 2021;8:103.
27. Rong Z, Shu X, Xu Z. Compact fiber curvature and temperature sensor inscribed by femtosecond laser through the coating. J Lightwave Technol. 2021;39:3981–90.
28. Salman AM, Al-Hayali SK, Al-Janabi A. Sensitivity-enhanced moisture sensor based on θ-shape bending fiber coated with copper-polyvinyl alcohol thin film. IEEE Sens J. 2021;21:21546–54.
29. Shao M, Liang J, Gao H, Wang C, Liu Y. Seven-core fiber based in-fiber Mach-Zehnder interferometer for temperature-immune curvature sensing. Optics Commun. 2023;528: 128989.
30. Shao M, Sun H, Zhang R, Li L, Liu Y, Qiao X. In-fiber humidity sensor based on Black Phosphorus-Polyvinyl alcohol. Opt Fiber Technol. 2022;68: 102782.
31. Su B, Zhang F, Zhong L, Qi B, Xu O, Qin Y. Temperature compensated curvature sensor with insensitive axial strain based on tapered ring core fiber interferometer. Optics Commun. 2022;513: 128067.
32. Sun X, Du H, Dong X, Hu Y, Duan JA. Simultaneous curvature and temperature sensing based on a novel mach-zehnder interferometer. Photonic Sensors. 2019;10:171–180.
33. Tian K, Zhang M, Farrell G, Wang R, Lewis E, Wang P. Highly sensitive strain sensor based on composite interference established within S-tapered multimode fiber structure. Opt Express. 2018;26:33982–92.
34. Tong Z, Zhao Y, Wang X, Li P, Zhang W, Zhang J. Research on dual-parameter biosensor based on no-core fiber coated by composite film. Optik. 2022;259: 169027.
35. Wang D, Jiang Y, Geng X, Yang B, Li L. Study of asymmetric biconical fiber tapers for in-fiber Mach-Zehnder interferometers and applications in single-frequency fiber lasers. Opt Express. 2021;29:14384–93.
36. Wang F, Lu Y, Wang X, Ma T, Li L, Yu K, Liu Y, Li C, Chen Y. A highly sensitive temperature sensor with a PDMS-coated tapered dispersion compensation fiber structure. Optics Commun. 2021;497: 127183.
37. Wang J, Wang A, Chen X, Liu S, Xu X, Sun C, Yan Y, Yan Q, Wang S, Geng T, et al. An all fiber mach-zehnder interferometer based on tapering core-offset joint for strain sensing. IEEE Photonics Technol Lett. 2022;34:11–4.
38. Wang S, Ma Y, Chen W, Wang S, Yi Y, Li X, Lu C, Zhang S, Geng T, Sun W, et al. Ultrasensitive strain sensor based on mach-zehnder interferometer with bent structures. J Lightwave Technol. 2021;39:6958–67.
39. Wang Y, Zhou Y, Liu Z, Chen D, Lu C, Tam HY. Sensitive Mach-Zehnder interferometric sensor based on a grapefruit microstructured fiber by lateral offset splicing. Opt Express. 2020;28:26564–71.
40. Xia F, Zhao Y, Peng Y. In-line microfiber MZI operating at two sides of the dispersion turning point for ultrasensitive RI and temperature measurement. Sens Actuators, A. 2020;301: 111754.
41. Xiao D, Wang G, Yu F, Liu S, Xu W, Shao L, Wang C, Fu H, Fu S, Shum PP, et al. Optical curvature sensor with high resolution based on in-line fiber Mach-Zehnder interferometer and microwave photonic filter. Opt Express. 2022;30:5402–13.
42. Xiao SY, Wu BL, Wang ZX, Jiang YC. A peanut taper based Mach-Zehnder interferometric sensor for strain and temperature discrimination. Optical Fiber Technol. 2022b;70.
43. Zhang F, Xu X, He J, Du B, Wang Y. Highly sensitive temperature sensor based on a polymer-infiltrated Mach-Zehnder interferometer created in graded index fiber. Opt Lett. 2019;44:2466–9.
44. Zhang G, Ge Q, Wang H, Wu X, Yu B. Stress applying TPMF based sensor for simultaneous temperature and strain measurement. Optik. 2022;254: 168636.

45. Zhang H, Gao S, Luo Y, Chen Z, Xiong S, Wan L, Huang X, Huang B, Feng Y, He M et al. Ultra-sensitive mach-zehnder interferometric temperature sensor based on liquid-filled D-shaped fiber cavity. Sensors (Basel). 2018;18.
46. Zhao C, Qiu H, Chen H, Hu X, Yu Q, Lian Z, Li J, Qu H. In-fiber Mach-Zehnder temperature sensor using silicone-oil-filled dual core fiber. Sens Actuators, A. 2021;323: 112644.
47. Zhao R, Shu X. Curvature sensor based on femtosecond laser-inscribed straight waveguide in FMF. Opt Laser Technol. 2022;152: 108154.
48. Zhao R, Shu X, Wang P. High-performance bending sensor based on femtosecond laser-inscribed in-fiber mach-zehnder interferometer. J Lightwave Technol. 2020;38:6371–8.
49. Zhao Y, Zhao J, Peng Y, Tong R-J, Cai L. Simultaneous measurement of seawater salinity and temperature with composite fiber-optic interferometer. IEEE Trans Instrum Meas. 2022;71:1–8.
50. Zhao Y, Zhao J, Wang X-X, Peng Y, Hu X-G. Femtosecond laser-inscribed fiber-optic sensor for seawater salinity and temperature measurements. Sens Actuators, B Chem. 2022;353: 131134.
51. Zhou X, Li S, Li X, Yan X, Zhang X, Cheng T. A vectorial analysis of the curvature sensor based on a dual-core photonic crystal fiber. IEEE Trans Instrum Meas. 2020;69:6564–70.
52. Zhou Z, Gong Z, Dong B, Ruan S, Chan CC. Enhanced sensitivity refractometer based on spherical mach-zehnder interferometer with side-polished structure. IEEE Sens J. 2021;21:1548–53.
53. Zhu X, Pan Y, Sun A, Sun D, Liu W, Cao J, Zhang G, Han Z, Zou L, Liang Z, et al. High sensitivity curvature sensor based on a double-sphere tapered no-core fiber Mach-Zehnder interferometer. Opt Laser Technol. 2022;155: 108364.
54. Zuo B-Z, Liang X, Zhang X-R, Yin T-H. Detection of refractive index with a temperature-compensated MZI-based optical sensor using few-mode fiber. IEEE Access. 2021;9:158651–9.

Chapter 4
Beyond the Spectrum: Specialty Optical Fibers in Magnetic Field Sensing

Jing Zhang, Longfei Liu, Zeyu Tan, Yongkang Huang, Zhichao Wu, and Tianye Huang

Abstract This chapter explores the advancements in magnetic field sensors based on optical fibers and advanced multi-material fibers. Traditional optical fibers face challenges in directly sensing magnetic fields due to their insulating properties against electromagnetic interference. To overcome this limitation, transduction materials and modified structures are employed, enabling the conversion of magnetic fields into measurable changes such as optical fiber's stress or refractive index. The chapter delves into magneto-active materials like magneto-fluidic, magneto-strictive, and magneto-optical materials. It also discusses the structures for enhancing sensitivity, including micron fibers, optical fiber gratings, and interferometers. Types of fiber-based magnetometers have been extended to multi-material multi-functional fibers with recent advances, particularly those drawn through thermal drawing processes. Novel designs to integrate magneto-active material and commercial microchips within the silica and polymer fibers, showcase potential advancements in distributed sensing. The chapter presents notable studies, demonstrating diverse applications such as high-sensitivity magnetic field imaging, environmental monitoring for geophysical exploration, and industrial applications in equipment

J. Zhang (✉) · Z. Tan · Y. Huang · Z. Wu · T. Huang
School of Mechanical Engineering and Electronic Information, China University of Geosciences (Wuhan), Wuhan 430074, China
e-mail: zhangjing28@cug.edu.cn

Z. Tan
e-mail: tzy1102597659@cug.edu.cn

Y. Huang
e-mail: huangyongkang@cug.edu.cn

Z. Wu
e-mail: wuzhichao@cug.edu.cn

T. Huang
e-mail: tianye_huang@163.com

J. Zhang · L. Liu
School of Future Technology, China University of Geosciences (Wuhan), Wuhan 430074, China
e-mail: liulongfei@cug.edu.cn

L. Wei (ed.), *Advanced Optical and Optoelectronic Fibers*, Advances in Optics and Optoelectronics, https://doi.org/10.1007/978-981-97-6218-7_4

monitoring. The evolving landscape of fiber-shaped magnetic field sensors promises enhanced capabilities and broader applications across industries.

1 Introduction

Magnetic field sensing plays a crucial role in various fields such as biomedical applications, military defense, resource exploration, deep Earth measurements, as well as our daily live matters. Magnetic field sensors can measure the characteristics of the magnetic field, like the strength, direction, and variation. Fiber optic sensing systems typically utilize optical fibers as the sensing mediums, which can sense acoustic waves, temperature, and stress through the intensity, phase, and polarization of light [1–3]. In recent years, single-point and distributed fiber optic systems have demonstrated significant potential in applications such as vibration sensing, temperature measurement, and strain monitoring. They have provided valuable assistance in the health monitoring of large structures (buildings, bridges, dams), deep Earth exploration, and oil and gas resource exploration [4–7].

Fiber optic magnetic field sensing is an important and popular research direction. Conventional optical fibers for communication are usually composed of materials like silica and polymers. These insulating materials naturally shield against electromagnetic interference, protecting the transmission of optical signals. However, this characteristic hinders the optical fiber's ability to sense magnetic fields. Consequently, fiber optic systems cannot directly measure magnetic fields, other transduction materials or structures are typically required to convert the magnetic field changes into stress or refractive index changes of fiber materials, that an optical fiber system can measure [8].

Due to this limitation, magnetic field sensors developed from traditional optical fibers are generally operated as single-point sensors [8–11]. The development of multi-material, multi-functional fibers based on the thermal-drawing process has provided new ideas for distributed magnetic field optical fiber sensors [12]. In the past, the thermal drawing process could only fabricate optical fibers with limited types of material, such as silica and polymer. With the development of materials science and the optical fiber industry, more types of materials, including semiconductors, metals, and functional polymers can also be drawn into fibers. At the same time, through the design and arrangement of these materials, these new optical fibers can serve as optical waveguides, but also provide more special functions, which are also known as multi-material multi-functional fiber (MMF) [13–16] Based on the thermal drawing method, functional materials that can sense magnetic fields now can be co-drawn with conventional optical fiber materials to form specialty optical fibers with magnetic sensitivity [17].

In this chapter, we will discuss magnetic field sensors based on traditional optical fibers, including those with embedded magnetic field transducer materials and sensitive-enhancing structures. We will also discuss the magnetic field sensors based on advanced multi-material multi-functional fibers. Meanwhile, we will briefly introduce the applications of these fiber-based magnetic field sensors.

2 Specialty Optical Fibers in Magnetic Field Sensing

- Working Mechanism of Optical Fiber-based Magnetometer

From the time optical fibers were invented, optical fibers were made of quartz glass and doped glasses for communication purposes. Subsequently, several plastic optical fibers have been developed [18]. These plastic fibers use polystyrene (PS), polymethylmethacrylate (PMMA), and polycarbonate (PC) as the core material, and PMMA and fluoroplastics as the cladding material. These materials are insulators, showing excellent properties in electromagnetic signal shielding. For optical signal transmission, the shielding property against electromagnetic interference is highly welcome since it protects the transmitted optical signal from influences of surrounding electromagnetic signals, and enhances the transmission accuracy. However, when using optical fiber for magnetic field sensing, the shielding property of optical fiber materials makes it difficult to sense magnetic fields.

How to make use of optical fibers, which are insensitive to magnetic fields, to achieve the perception of magnetic field signals is a key question awaiting answers in this research direction. By choosing suitable sensing materials and structures, the magnetic field information can be transduced into readable signals such as stress and refractive index changes, which can be accurately sensed by optical fibers.

– Magneto-active Materials

Magneto-active materials can convert magnetic fields into other forms of energy, working as magnetic transducers. For developing fiber optic sensing systems, there are three commonly used magnetically sensitive materials: magneto-fluid (MF), magneto-strictive (MS), and magneto-optical (MO) materials. These materials can convert the directional strength and change of the magnetic field into the change of refractive index of the optical fibers materials or polarization of light.

A common class of magnetic fluid materials is ferromagnetic fluids, which are made of nanoscale ferromagnetic or ferrimagnetic particles (e.g., γ-Fe_2O_3, Fe_3O_4, α-Fe_3N) suspended in a carrier fluid, forming a separate clump of magnetic dust. Without a magnetic field, the magnetic nanoparticles form a disordered distribution. During the process of magnetization, the magnetic force will guide these magnetic nanoparticles to reconfigure their orientation and link from a disordered state to an ordered state (Fig. 1a), which changes the effective refractive index of MF. The stronger the magnetic field, the longer the chain formed by the magnetic particles. The refractive index of the MF changes during the reconfigurations of particle chains, which the optical fiber sensing system can detect [19].

Magneto-strictive (MS) materials are another commonly used transducer material for optical fiber magnetic field sensors. MS materials can change their shapes and

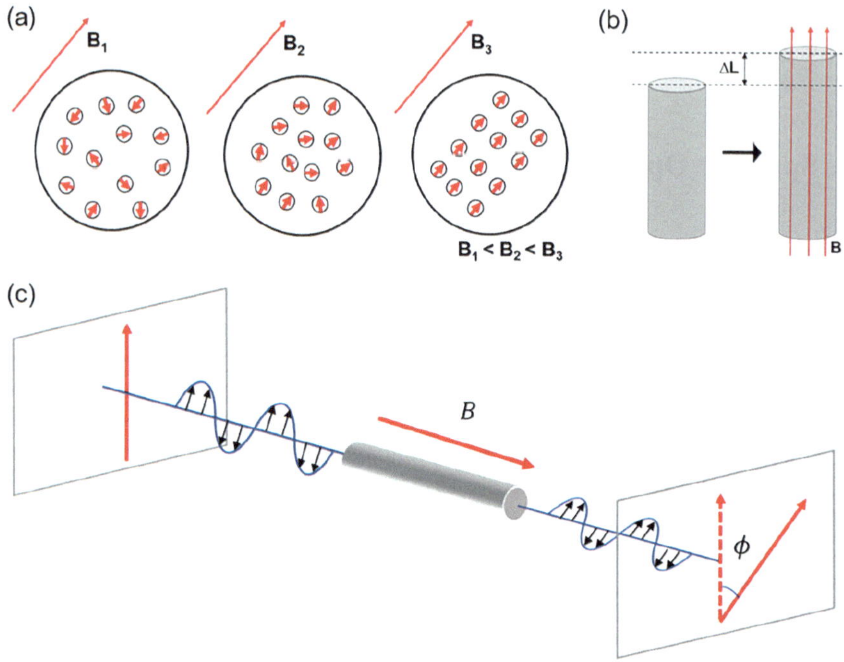

Fig. 1 **a** Variation of magnetic fluid nanoparticles under different magnetic field strengths. (Magnetic field strength: $B_3 > B_2 > B_1$). **b** Deformation of magneto-strictive materials under magnetic fields. **c** The principle of magneto-optical effect

dimensions during the process of magnetization. The amount of extension and shortening of magneto-strictive material changes depending on the magnetic field strength, as well as the magneto-strictive coefficient of the material (Fig. 1b). Through effective coupling, magneto-strictive materials can convert the changes in a magnetic field into a length change of the optical fiber, which can generate stresses and strains in the fiber, and then be detected by the fiber. Commonly used magneto-strictive materials in fiber optic sensing systems include Terfenol-D ($Tb_xDy_{1-x}Fe_2$), and nickel–iron alloys with a large magnetostriction force [20].

The magneto-optical (MO) effect refers to the phenomenon that when a beam of polarized light passes through a magneto-optical material as shown in Fig. 1c, the magnetic field in the direction of the beam's propagation causes the polarization plane of the polarized light to rotate. The MO materials can be drawn or added into optical fibers, and the direction of polarization of light propagating in these specially modified optical fibers can be changed by the external magnetic field.

– Sensitivity-enhancing Structures

As the magneto-active material transduces the changes of a magnetic field into the changes of refractive index or stress of optical fibers, the structures of optical fiber shall also be adjusted accordingly to increase its sensitivity. Commonly used

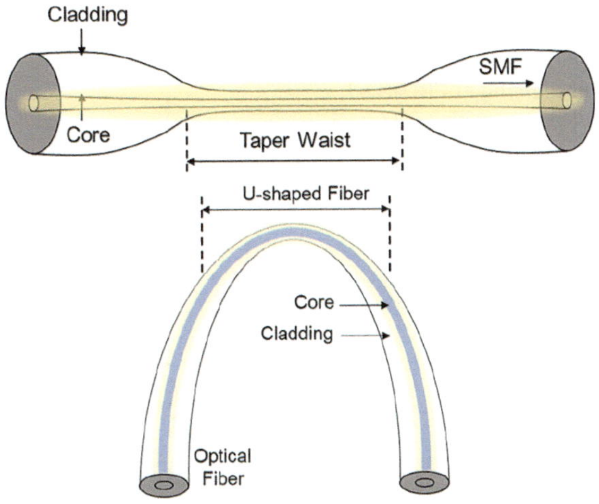

Fig. 2 The schematic diagrams of tapered micro-fiber (upper) and reshaped U-shape fiber (down)

sensitivity-enhancing structures for optical fiber magnetic sensors are micron fiber with evanescent field, optical fiber grating, optical fiber interferometer structures, etc.

Optical fiber is a very ideal optical waveguide, where light waves propagate forward in the fiber core through the phenomenon of total reflection, only a very small amount of light energy is transmitted into the fiber cladding, forming the evanescent light waves. Normally, the evanescent light is dissipated in the fiber cladding and does not interact with external subjects. By reducing the diameter of the fiber (fiber tapering, fiber U-shaping), or removing part of the cladding (fiber side polishing), the evanescent light can be exposed as shown in Fig. 2 [21]. Then the evanescent light wave can react with surrounding magnetic sensitivity materials, such as magnetic fluid or magneto-strictive material film.

Optical fiber grating is another sensitivity-enhancing structure with high efficiency. The fiber Bragg grating (FBG), long-period fiber grating (LPFG), and tilted fiber grating (TFG) are the most used fiber gratings (Fig. 3). An FBG is a typical type of distributed Bragg reflector written in a short segment of optical fiber, which reflects lights of selected wavelengths and transmits all others [22]. Light with the wavelengths that satisfy the Bragg condition will be strongly reflected. The reflected wavelength can be calculated by $\lambda_{Bragg} = 2n\Lambda,$ where λ_{Bragg} is the reflected wavelength, n is the refractive index of the fiber core and Λ is the grating period. n and Λ are both affected by temperature as well as stress. So, the reflected wavelength is also affected by changes in temperature and stress. When the grating period is greater than the wavelength of light propagating in the fiber, it becomes an LPFG, which couples the co-directionally propagated core mode and cladding mode, creating a series of loss peaks in the transmission spectrum. The changes in the effective refractive index of the cladding mode affect the displacement of the LPFG resonance peaks. Tilted fiber grating (TFG) is a special grating whose grating planes are slanted to the fiber axis. These tilts break the symmetry of the fiber, enhancing the light coupling from

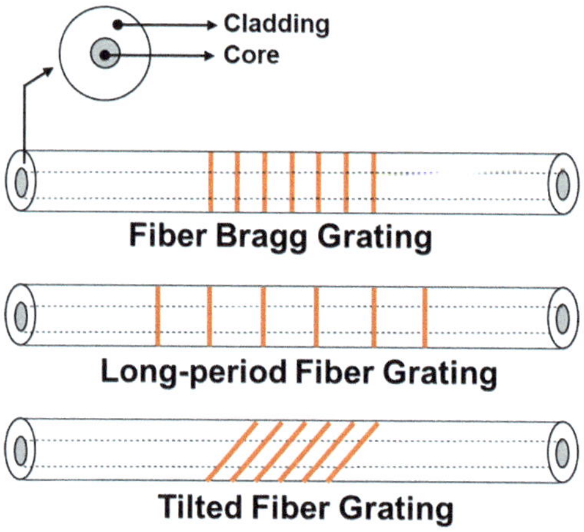

Fig. 3 The schematic diagrams of fiber Bragg grating (FBG), long-period fiber grating (LFBG), and tiled fiber grating (TFG)

the fiber core to the cladding. The cladding modes are susceptible to factors such as the refractive index of surroundings, which in turn increases the sensitivity of fiber sensing.

In addition, interferometers that use the superposition of light waves to obtain the sensing of small changes in phase and intensity of the wave, are commonly used in precision measurements. Gravitational waves, for example, are detected by a large-scale Michelson interferometer [23]. Interferometers are also commonly used in optical fiber sensing systems, including the Fabry–Perot interferometer (FPI) (Fig. 4a), the Michelson interferometer (MI) (Fig. 4b), and the Mach–Zehnder interferometer (MZI) (Fig. 4c). The principle of FPI is the superposition of multiple reflections of light between two reflective surfaces in a cavity, and the change of refractive index of the medium in the cavity or the distance between the two reflective surfaces will lead to the change of the interference wavelength. The FPI-based fiber magnetic sensors are usually filled with magnetic material inside the cavity to obtain information about the magnetic field by demodulation of interference wavelength changes. MI divides the source light into two beams that differ by 180°, one of which is reflected to the receiver by a fixed reflector, and the other is reflected to the receiver by a movable reflector to form an interference pattern. When the magneto-active materials and the moving reflector are combined, the magnetic field changes the reflector position thus changing the optical range difference as well as the interference pattern. MZI-based fiber sensors typically have two optical paths, one path for reference and the other for sensing [24]. By combining the sensing optical path with magnetic sensing materials, the magnetic field changes the phase of the optical signal. The relative phase shift change of the two optical paths is converted into magnetic field information.

– Delving Deeper: Types of Optical Fiber Magnetic Sensors

In this section, we will introduce some types of optical fiber magnetic sensors that are constructed by matching the sensitivity-enhancing materials and structures. In

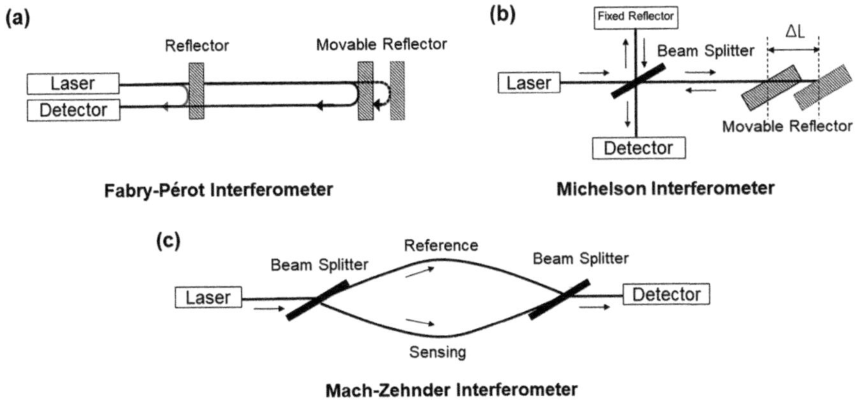

Fig. 4 The schematic diagrams of different interferometers **a** Fabry-Pérot interferometer. **b** Michelson interferometer. **c** Mach–Zehnder interferometer

2017. Herrera-Piad L A et al. designed a compact, magnetic field sensor system based on a short bi-tapered optical fiber (BTOF) span lying on a magnetic tape, which achieved a phase sensitivity of around 0.28 rad/mT [25]. In 2021, Nunzio Cennamo et al. presented a surface plasmon resonance (SPR) sensor based on a D-shaped plastic optical fiber (POF) and magnetic fluids (MFs). The sensor features a sensitivity of 6800 pm/mT over a range of characterized values between 0.15 and 1.2 mT [26]. In 2021, Liangquan Zhu et al. proposed a compact vector magnetometer based on a balloon-like fiber structure functionalized by magnetic fluid (MF). This sensor exhibits a high magnetic field intensity sensitivity of 0.78 nm/mT and a high magnetic-field direction sensitivity of 0.094°/mT [27].

In 2011, Ciming Zhou et al. etched the FBG into 22% HF solution and then packed the thinned FBG into the solution filled with magnetic fluid to make a magnetic field sensor. The magnetic field changed the refractive index of the MF, which in turn changed the reflection wavelength of the FBG. Experimental results showed that the sensor recorded a sensitivity of 2.3 pm/mT [28]. In 2022, Rahul Kumar proposed a magnetic field sensor based on a tapered small-core single-mode fiber structure, as shown in Fig. 5a, the tapered portion of the fiber was contained by magnetic fluid filled with ferromagnetic particles (FMPs) with varying concentrations. The magnetic field sensor had a maximum sensitivity of 0.46 nm/mT at a taper waist diameter of 10 μm and a tapered waist surrounded by MFs with 1.22% FMPS [29]. In 2020, Yaofei Chen et al. designed a vector magnetic field sensor based on a side-polished single-mode-multimode-single-mode fiber (Fig. 5b), magnetic fluid was wrapped around the evanescent field at the side polish. It was found that reducing the residual thickness (RT) of the side-polished multimode fiber enhanced the sensitivity to magnetic field strength and direction, which were −0.053 nm/Oe and −5.68 nm/degree, respectively, when the RT was 50.5 μm [30].

In 2021, Liangquan Zhu et al. proposed a vector magnetic field sensor based on the magnetic fluid single-mode fiber structure. The single-mode optical fiber was

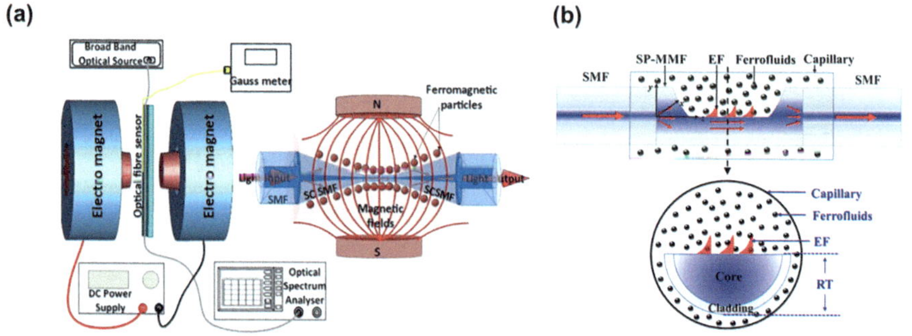

Fig. 5 **a** Tapered small-core single-mode optical fiber sensor surrounded by magnetic fluid [29]. Copyright 2022, MDPI. Reproduced with permission. **b** Magnetic field sensor based on side polished optical fiber MZI and MF [30]. Copyright 2022, IEEE Xplore. Reproduced with permission

bent into a spherical structure and the shape was then fixed on alcohol flame. Due to the bending, the core mode of the fiber leaks into the cladding, resulting in a higher-order cladding mode that is sensitive to external refractive index changes. The sensor achieved a high magnetic field strength sensitivity of 0.78 nm/mT and a high magnetic field direction sensitivity of 0.094°/mT [27].

In 2015, Nancy Meng Ying Zhang et al. proposed a magnetic field sensor with MF coated on a long-period grating (Fig. 6). The sensor exhibited a sensitivity of 0.154 dB/Gauss at 7.4 Gauss [31]. In the same year, Yinping Miao et al. proposed a magnetic field sensor based on MF and a tilted long-period grating (TLPFG). The TLPFG had a grating period of 650 μm and the TLPFG is manufactured by high-frequency CO_2 laser pulses and sealed with MF. The sensor achieved a sensitivity of 0.05 dB/Oe in the range of 75 Oe to 300 Oe [32].

In 2022, Shuhao Gai et al. proposed a highly sensitive vector magnetic field sensor based on optical fibers and MZI. A microcavity was fabricated on an optical fiber by a laser-induced micro-plasma process, which acts as the fiber MZI (FMZI), I_1, and I_2 in Fig. 7a represent the sensing optical path and the reference optical path, respectively. The microcavity FMZI is combined with a magnetic fluid to form a

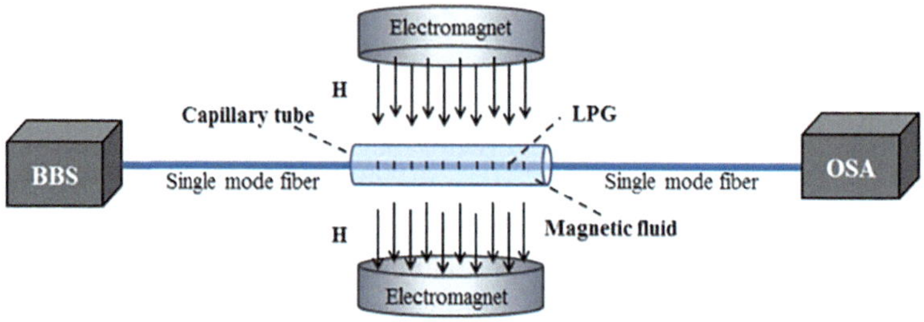

Fig. 6 Schematic of the magnetic field sensor based on MF and long-period gratings [31]. Copyright 2015, IOP Publishing. Reproduction with permission

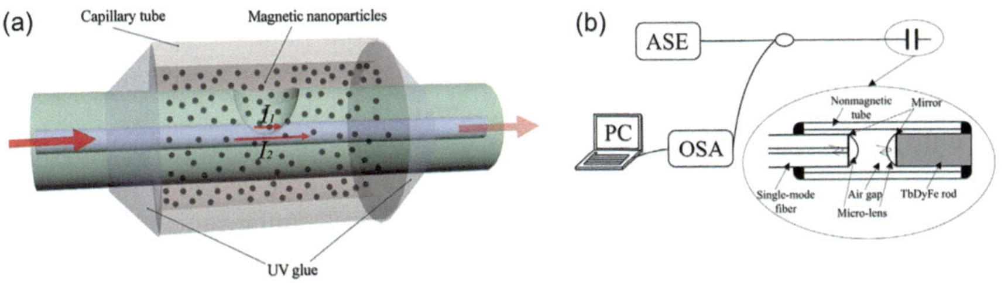

Fig. 7 **a** Schematic of the micro-cavity of the MZI-based sensor [33]. Copyright 2022, Elsevier. Reproduction with permission. **b** Schematic diagram of a magnetic field sensor based on an FP cavity and the TbDyFe rod [34]. Copyright 2015, Elsevier. Reproduction with permission

vector magnetic field sensor. The sensor achieved the maximum intensity sensitivity of 64 nm/mT [33].

In 2014, Riqing Lv et al. proposed a magnetic field sensor based on MF and FPI. The MF was filled into the FP cavity and the effective refractive index change of MF was measured to achieve magnetic sensing. The sensor achieved a sensitivity of 0.0431 nm/Gs at 0–400 Gs with a resolution higher than 0.5 Gs [35]. In 2015, Feifei Chen et al. designed a weak magnetic field sensor based on Fabry–Perot interference. The FP cavity consists of an optical fiber end-face and a TbDyFe rod end-face. Each end-face is covered by a mirror with micro-lens (Fig. 7b). DC magnetic field variation causes a change in the length of the TbDyFe rod, which changes the length of the FP cavity. The sensor demonstrated a high sensitivity of 1510 nm/mT and a magnetic resolution of 25 nT [34]. In Table 1, we list some properties of typical fiber optic magnetic field sensors, including sensing structure, materials, sensitivity, and magnetic field measurement ranges.

- Working Mechanism of Multi-material Multi-function Fiber Based Magnetometers.

Existing optical fiber-based magnetic field sensors require a well-defined fiber structure and magnetic transducer material covering. The fabrication process is complex, and their sensing accuracy leaves plenty of room to be improved. Furthermore,

Table 1 Technical parameters of fiber optic magnetic field sensor

Sensing Structure	Materials	Sensitivity	Measured Range	References
FBG	Terfenol-D	9.83 pm/mT	12–150 mT	[36]
FBG	Terfenol-D	8.4 pm/mT	20–80 mT	[37]
MZI	MF	240 pm/mT	34–42 mT	[38]
Side-Polished Fiber	MF	6.92 pm/mT	2–16 mT	[39]
FBG	MF	1.159 pm/mT	0–18 mT	[40]
FPI	Terfenol-D	101 pm/mT	10–30 mT	[41]
Photonic Crystal Fiber	MF	650 pm/mT	5–13 mT	[42]

multiple processing steps and modification schemes at present can only meet the demand for single-point sensing. New technologies are strongly needed for highly sensitive, distributed optical fiber-based magnetic field sensors.

With the development of the optical fiber industry and material science, new types of multi-material multifunctional optical fiber can be prepared by the fiber thermal drawing process. Traditional optical fibers for communication are generally composed of a single material such as quartz glass or polymer, whose refractive index can be modified through material doping, forming an optical waveguide. Now, with the thermal drawing process, the materials for fabricating optical fibers are not limited to silica or polymer, and the fiber structures are not limited to the core-cladding structures as well.

The selection of materials that can be drawn into fibers now has been expanded to semiconductors (silicon, germanium, gallium antimonide, etc.), metals (gold, silver, platinum, copper), and functional polymers. The structures of the drawn fiber are in microns, but the structures can be precisely designed on a macro-scale by constructing the fiber preform before drawing. Then the thermal drawing process will scale the designed structure down to micro-scale to form a fiber. The novel advanced fibers are called multi-material multi-functional fibers (MMFs). To date, multi-functional multi-material optical fibers (MMF) with different dielectric, optical, photoelectric, magnetic, acoustic, mechanical, electronic, and drug transport functions have been developed and achieved remarkable accomplishments in the fields of photoelectric detection, temperature sensing, pressure sensing, photoelectric detection, chemical detection, laser modulation, micro and nanomanufacturing, magnetic detection, and smart fabrics.

In a way, sensitivity-enhancing structures and materials can be integrated into the fiber directly through the thermal drawn, eliminating the problems in multiple-step processing and alignment. Here we present some new fiber structures and principles for distributed measurements of magnetic field.

- Thermally Co-drawn Advanced Magnetic/Glass Multi-Material Fibers

When fabricating MMFs, it is necessary to consider that the melting/softening temperatures of selected materials, shall be similar for simultaneous processing, and the coefficients of thermal expansion of these materials should be comparable as well to avoid stress damage. In addition, the cladding material is needed to provide mechanical support for the whole structure during the thermal drawing process. Considering these limitations, some magneto-strictive materials are selected to be drawn into fibers together with glass materials.

In 2014, Daniel Homa et al. achieved a series of advanced multi-material fibers, verifying the possibility of co-drawing the silica with magnetic-sensitive materials. The magnetic sensitive fibers take Metglas and Terfenol-D as the fiber cores and silica glass as the fiber cladding (as shown in Figs. 8a, 8b). It is noteworthy that the metallic glass core fiber retains its magnetic-sensitive properties after fiber thermal drawing [43]. In 2017, Guowu Tang et al. proposed a glass-cladding Bi core fiber, as shown in Fig. 8c. The fiber was also fabricated by the molten core thermal drawing method, and the core diameter was kept at 100 μm to 150 μm with a total fiber

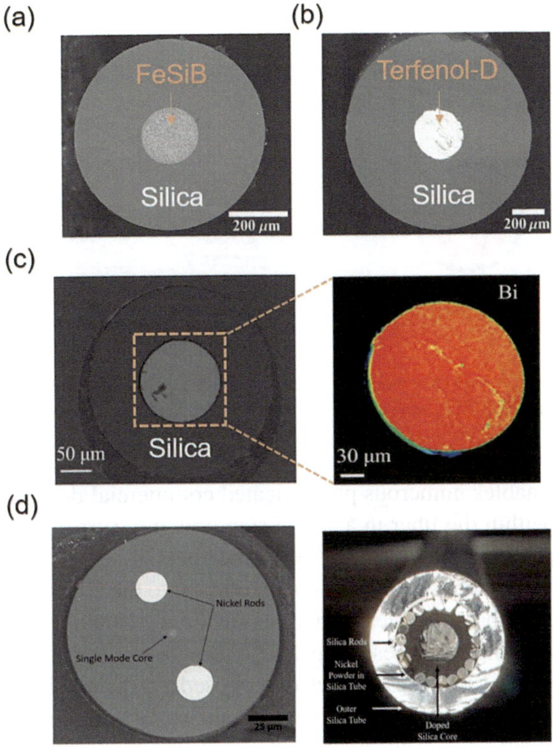

Fig. 8 Cross sections of magneto-strictive material and silica cladding fibers. **a** Scanning electron microscopy images of **a** fibers with FeSiB core diameter of about 110 μm and fused silica cladding of 630 μm.**b** Fibers with Terfenol-D core diameter of about 220 μm and fused silica cladding of 950 μm [43]. Copyright 2014, Elsevier. Reproduction with permission. **c** Bi-core fiber structure schematic, and EPMA images of fiber core cross-section [44]. Copyright 2017, Elsevier. Reproduction with permission. **d** The optical fiber with nickel rods, and the photo of the fiber preforms [45]. Copyright 2022, IEEE Xplore. Reproduction with permission

diameter of 300 μm to 450 μm. Since Bi is a highly magneto-resistive material, this fiber became a great substance for fiber optic magnetic field sensing [44].

In 2022, Zachary Hileman et al. proposed a design of multi-material fiber for optic magnetic sensing, which was based on the magneto-strictive effect. The fiber can be developed into a distributed sensing system. The sensor system has an optical fiber core-cladding structure, consisting of a silica core (as shown in Fig. 8d), silica cladding, and two ferromagnetic nickel rods. To enhance the sensitivity of the fiber, a series of fiber Bragg gratings were written. The sensor demonstrated a linear response of 0.1 εμ/μT, which was further enhanced in subsequent studies using Metglas or Terfenol-D instead of Ni [45].

– Thermally Drawn Diode Fibers Enabling Magnetic Field Sensing

With the development of the thermal drawing process, not only raw materials are drawn into optical and multifunctional fibers, but also artificial microscale devices can be drawn into the fibers. Initially, discrete devices are incorporated into the preform without establishing connections. Subsequently, electrical wires are introduced into the milled channels within the preform during the drawing process. As the preform is downscaled into a smaller microscale fiber, the wires gradually converge until they contact the devices, which is termed the "convergence" process. This method

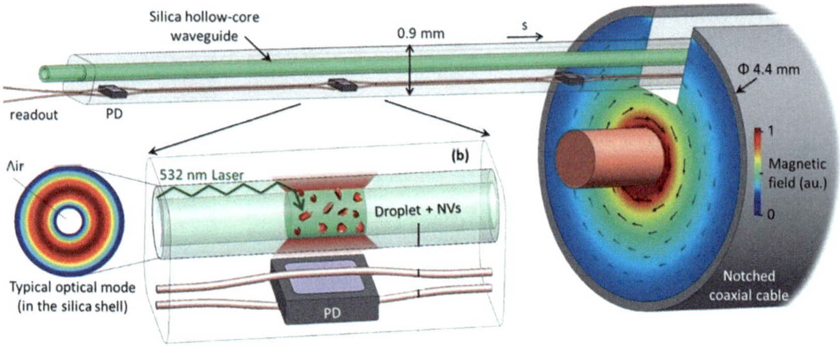

Fig. 9 MMF sensing structure with microchips allows distributed long-distance magnetic measurements [46]. Copyright 2019, WileyVCH. Reproduction with permission

enables numerous prefabricated commercial devices to establish electrical contacts within the fiber in a single-step drawing, surpassing the performance of previously thermally drawn multifunctional fibers. Indeed, this method provides a "gateway" for incorporating semiconductor chips into the advanced multifunctional fiber, enabling the fibers to access functions such as computation, illumination, and optoelectronic sensing with these chips.

In 2019, Yoel Fink et al. proposed a distributed optical fiber magnetic field magnetometer based on a nitrogen-vacancy (NV) quantum. The structure enables distributed magnetic sensing over a long length. As shown in Fig. 9, the key to achieve this structure was by placing hundreds of parallel embedded photodiodes, two electric wires, and a hollow optical waveguide into the polymer cladded fiber through the thermal drawing. The hollow optical waveguide then was filled with a few drops of NV diamond fluid. The fiber is placed in a larger coaxial cable that provides the required excitation. The distributed sensor demonstrated a total length of 90 m and features 102 detection points. The sensitivity of the sensing system was 63 ± 5 nT/$\sqrt{(\text{Hz})}$ per detection point [46].

- Applications of fiber-based magnetic field sensors

– Quasi-distributed and Distributed Sensing System

Fiber-optic magnetic field sensors have been comprehensively studied and developed in various fields and industries because of their small size, high sensitivity, long-distance transmission, and hybrid sensing capability. As introduced in the previous sections, the fiber optic single-point magnetic field sensors have been well developed with high accuracy. However, single-point sensors cannot meet the needs of all scenarios, and the use of distributed sensors is strongly required including rail transportation monitoring, transmission line monitoring, large construction quality monitoring, geomagnetic measurement, etc. For this reason, several distributed or quasi-distributed systems have been developed, which will be discussed in this section.

In 1993, an interferometric fiber optic sensor system was introduced by F. Bucholtz et al. The system consisted of two triaxial fiber optic magnetometers which were placed on the sea floor and operated remotely via fiber optic cables, as shown in Fig. 10a. Each seafloor node contained a single-mode fiber optic Michelson interferometer with three magneto-strictive sensors and two piezoelectric elements. The results of the first deployment of the system were relatively good for subsea operations. The system obtained an interferometer phase resolution of 0.6 μrad/$\sqrt{}$(Hz) and a magnetic field resolution of at least 150 pT/$\sqrt{}$(Hz) at 0.1 Hz on all axes [47]. Then the system was placed on the seafloor for one year of operation, and the entire sensor array consisted of eight interferometers, within one individual interferometer (Fig. 10b). Three components of the Earth's magnetic field were measured simultaneously with the fiber-optic magnetometer array and the shore-based fluxgate magnetometer over 48 h, as shown in Fig. 10c. The measured values of the system and the tri-axial fluxgate magnetometer demonstrated a strong correlation. The system was capable of operating unattended for several months, excluding data storage and power outage limitations [48].

In 2022, Yiwen Ou et al. proposed and demonstrated a quasi-distributed fiber optic magnetic field sensing system based on the frequency-shifted interferometric

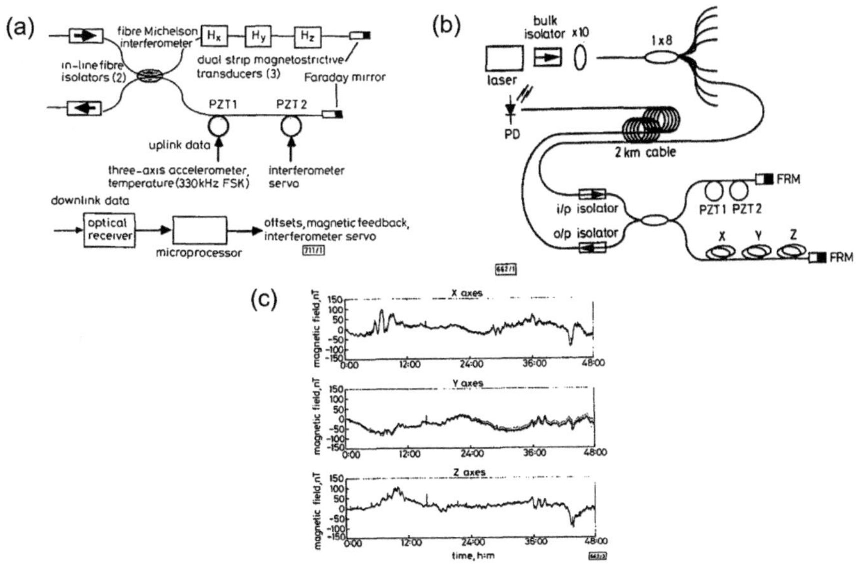

Fig. 10 **a** Fiber optic magnetic sensing consists of two triaxial fiber optic magnetometers [47]. Copyright 1993, Wiley. Reprinted with permission. **b** System diagram of sensor arrays deployed on the seabed. **c** Changes in the three components of the Earth's magnetic field were measured simultaneously with a fiber optic magnetometer array and a shore-based fluxgate magnetometer over 48 h [48]. Copyright 1994, Wiley. Reprinted with permission

fiber cavity oscillation (FSI-FCRD) technique. In this system, as shown in Fig. 11, multiple sensing units are connected in series. In each unit, a laterally polished single-mode fiber is coated with a dilute water-based ferrofluid is used as the sensing element to improve sensitivity and mechanical strength. The cavity loss of each cell varies with the applied magnetic field, and the magnetic field can be detected by monitoring the cavity loss variation. The scheme was designed with several sensing units connected in series at 2500 m. The sensitivities obtained were 6.7894×10^{-3} and 7.4980×10^{-3} dB/mT, respectively. The scheme indicates that the maximum number of sensing units that can be multiplexed is up to 30 at a distance of 15.8 km in the same experimental setup, allowing for distributed magnetic field measurements over a large area [49].

- Hybrid Sensing Systems for Magnetic Field

Fiber optic sensing systems are now known for their ability to sense the temperature, vibrations/acoustic waves, and strain/stress. In many applications, not only the magnetic field, but also vibration, temperatures, and strains shall be recorded since they are all critical parameters for *in-situ* measurements. For example, in deep earth exploration, the simultaneous in-situ sensing of multiple geophysical fields, such as magnetic field, temperature, and vibration, can help to solve the problem of the multi-solution nature of geological structure inversion. Developing hybrid sensing systems generally faces two problems: what principles and structures can be employed for simultaneously sensing different parameters, and how the mixed sensing signals can be demodulated. For multi-parameter hybrid sensing, various types of resonant

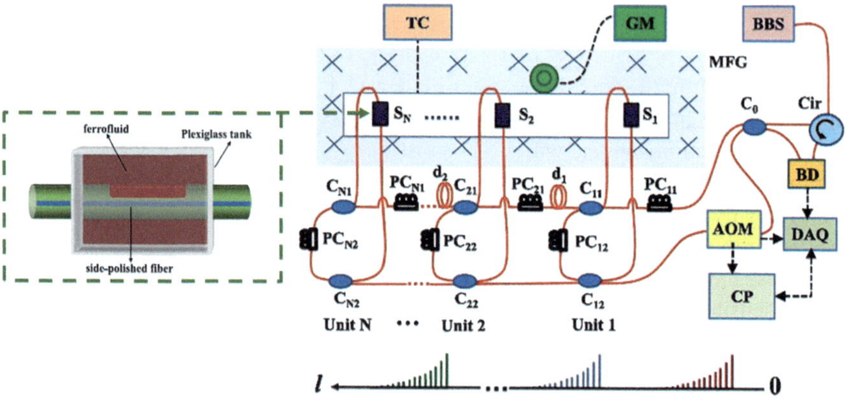

Fig. 11 Diagram of the experimental setup of a quasi-distributed magnetic field sensing system with multiple individual sensing units. TC: temperature controller; GM: Gauss meter; MFG: magnetic field generator; BBS: a broadband light source; Sn: the nth magnetic field sensing element; C_0: 3 dB fiber coupler; Cir: fiber circulator; C_{n1} and C_{n2}: the pair of 99.5/0.5 fiber couplers in the nth fiber RDC (the nth sensing unit, n = 1,2, ..., N); PC_{n1} and PC_{n2}: the pair of polarization controllers in the nth channel; di: the ith fiber delay line, i = 1,2,..., N–1; BD: balanced detector; AOM: acoustic-optic modulator; DAQ: data acquisition card; CP: computer [49]. Copyright 2022, Elsevier. Reprinted with permission

structures can be superimposed on the original fiber sensitivity-enhancing structures, achieving the hybrid sensing systems. For the demodulations, different demodulation algorithms can be used to solve these problems. This section briefly introduces some well-functioning fiber-based hybrid sensing systems for sensing magnetic fields and others.

In 2018, Xiaohong Bai et al. proposed an optical fiber sensor that simultaneously measures magnetic and acoustic fields. The sensor consisted of two cascaded fiber Bragg gratings, one of which was inserted into the TbDyFe rod for magnetic sensing, and another FBG was used for acoustic sensing. These two FBGs were fixed in a rectangular polymethyl methacrylate (PMMA) frame (Figure 12a). For acoustic wave detection, the sensor achieved a signal-to-noise ratio of 52.9 dB (@300 Hz) and 47.3 dB (@1 MHz). For magnetic field detection, the sensor achieved a sensitivity of 0.82 dB/mT between 1.45 mT–22.29 mT [50].

In 2018, Rui Ma et al. proposed a distributed feedback fiber laser sensor for measuring magnetic fields and acoustic waves (Fig. 12b). The fiber was held between a TbDyFe tube and a cover. The magnetic field causes a longitudinal deformation of the Terfenol-D tube, resulting in a change in the fiber length, which in turn causes a change in the resonant frequency of the fiber laser. In this way, the resonant frequency

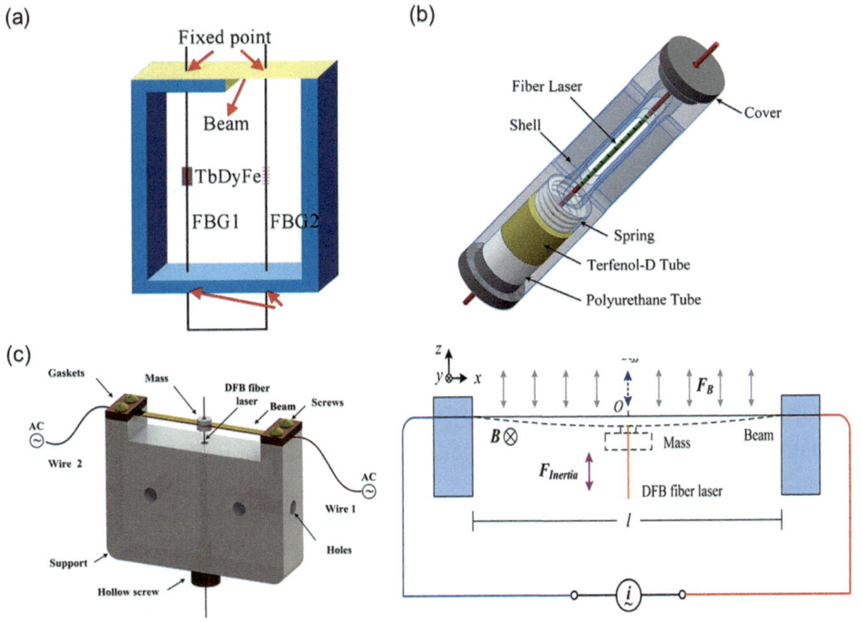

Fig. 12 **a** Structure of a sound and temperature sensor consisting of two fiber Bragg gratings in cascade [50]. Copyright 2018, Elsevier. Reproduction with permission. **b** Schematic of a distributed feedback (DFB) fiber laser sensor for simultaneous measurement of acoustic and magnetic fields [51]. Copyright 2018, OPG. Reproduction with permission. **c** Schematic diagram of an acceleration-magnetic sensor and the force diagram of the sensor [52]. Copyright 2017, OPG. Reproduction with permission

shifts of the fiber laser become an identification of the variation of the magnetic field. In another way, acoustic pressure can be measured by reading the phase shifts of the fiber laser. The two signals can be demodulated simultaneously in the frequency domain without affecting each other. Experimental studies have shown that the sensor demonstrated an acoustic pressure sensitivity of approximately 130 dB (0 dB re 1 pm/μPa) and a magnetic field sensitivity of 0.57 Hz/mT [51].

In 2017, Zhaogang Wang and colleagues introduced an acceleration-magnetometer utilizing a fiber laser for simultaneous measuring acceleration and magnetic fields (Fig. 12c on the left). The sensor itself, illustrated in Fig. 12c on the right, consists of a mass block and a beam, creating a spring-mass system along the z-direction. When inertial forces act upon the sensor, the mass block maintains its inertia while the beam undergoes deformation. This deformation induces strain on the distributed feedback (DFB) fiber laser. The displacement in the center wavelength of the DFB fiber laser is indicative of the acceleration. Simultaneously, when the sensor is exposed to a magnetic field, the metal beam, connected to a current source, generates an ampere force, causing a shift in the central wavelength of the DFB fiber laser. The sensor demonstrated an acceleration sensitivity of 9.75 pm/g within the 100–250 Hz range and a magnetic flux intensity sensitivity of 0.955 pm/T/mA at 725 Hz [52].

3 Conclusion

In conclusion, the exploration of optical fiber-based magnetic field sensors has unveiled a transformative landscape in sensing technologies. The integration of magnetic transducers (magneto-fluids, magneto-strictive materials, and magneto-optical materials) has enabled the transduction of magnetic field information into measurable changes in optical fiber properties. Sensitivity-enhancing structures, including micron fibers with evanescent fields, various optical fiber gratings, and interferometers, have been pivotal in enhancing the precision of magnetic field sensors.

Furthermore, the emergence of multi-material multi-functional fibers, facilitated by thermal drawing processes, has opened new frontiers in fiber-based distributed magnetic field sensing. Notable studies incorporating magneto-strictive materials, NV micro-diamonds, and photodetector chips into silica or polymer fibers present promising avenues for highly achieving sensitivity and distributed sensing systems. This technique of co-drawing magnetically sensitive materials with optical fibers overcomes the electromagnetic shielding of optical fibers and, from the basic structure and principles, gives optical fibers the ability to sense magnetic fields. The discussed applications, spanning high-sensitivity magnetic field imaging, environmental monitoring for geophysical exploration, and industrial equipment monitoring, underscore the versatility and broad utility of these sensors.

With the advancement of technology, continued developments in materials science and fiber optic techniques are poised to propel fiber optic magnetic field sensors to unprecedented heights. The integration of magnetically sensitive materials, sensitivity-enhancing structures, and commercial microscale chip devices into these fibers further enhances their performances. The potential for distributed magnetic field measurements over extensive areas showcases the scalability and adaptability of these sensors across diverse domains. The journey from traditional optical fibers to multi-material multi-functional fibers marks a deeper integration and shift of technology, promising a future where fiber optic magnetic field sensors play a pivotal role in shaping technologies of deep earth exploration and addressing multifaceted sensing needs.

References

1. Lee B. Review of the present status of optical fiber sensors. Opt Fiber Technol. 2003;9(2):57–79.
2. Lu P, Lalam N, Badar M, Liu B, Chorpening BT, Buric MP et al. Distributed optical fiber sensing: Review and perspective. Appl Phy Rev. 2019;6(4).
3. Pendão C, Silva I. Optical fiber sensors and sensing networks: Overview of the main principles and applications. Sensors. 2022;22(19):7554.
4. Barrias A, Casas JR, Villalba S. A review of distributed optical fiber sensors for civil engineering applications. Sensors. 2016;16(5):748.
5. García I, Zubia J, Durana G, Aldabaldetreku G, Illarramendi MA, Villatoro J. Optical fiber sensors for aircraft structural health monitoring. Sensors. 2015;15(7):15494–519.
6. Johny J, Amos S, Prabhu R. Optical fibre-based sensors for oil and gas applications. Sensors. 2021;21(18):6047.
7. Yang H, Peng J, Sun X, Chen G. Evaluation and research of comprehensive seismic observation technology based on optical fiber sensing. Arab J Geosci. 2021;14:1–13.
8. Liu C, Shen T, Wu H-B, Feng Y, Chen J-J. Applications of magneto-strictive, magneto-optical, magnetic fluid materials in optical fiber current sensors and optical fiber magnetic field sensors: a review. Opt Fiber Technol. 2021;65: 102634.
9. Alberto N, Domingues MF, Marques C, André P, Antunes P. Optical fiber magnetic field sensors based on magnetic fluid: a Review. Sensors. 2018;18(12):4325.
10. Sabri N, Aljunid S, Salim M, Fouad S. Fiber optic sensors: short review and applications. In recent trends in physics of material science and technology, by F. Gaol,K. Shrivastava,J. Akhtar (Springer, Singapore, 2015), p 299–311.
11. Zhang J, Wang C, Chen Y, Xiang Y, Huang T, Shum PP, et al. Fiber structures and material science in optical fiber magnetic field sensors. Front Optoelectron. 2022;15(1):34.
12. Hileman ZD. Fully Distributed multi-material magnetic sensing structures for multiparameter DAS applications. (Virginia Tech, Blacksburg, 2022).
13. Abouraddy AF, Bayindir M, Benoit G, Hart SD, Kuriki K, Orf N, et al. Towards multimaterial multifunctional fibres that see, hear, sense and communicate. Nat Mater. 2007;6(5):336–47.
14. Loke G, Yan W, Khudiyev T, Noel G, Fink Y. Recent progress and perspectives of thermally drawn multimaterial fiber electronics. Adv Mater. 2020;32(1):1904911.
15. Yan W, Page A, Nguyen-Dang T, Qu Y, Sordo F, Wei L, et al. Advanced multimaterial electronic and optoelectronic fibers and textiles. Adv Mater. 2019;31(1):1802348.
16. Zhang J, Wang Z, Wang Z, Wei L. Advanced multi-material optoelectronic fibers: a review. J Light Technol. 2021;39(12):3836–45.
17. Shen Y, Wang Z, Wang Z, Wang J, Yang X, Zheng X, et al. Thermally drawn multifunctional fibers: toward the next generation of information technology. InfoMat. 2022;4(7): e12318.
18. Kao CK. Optical fiber systems: technology, design, and applications. New York: McGraw-Hill; 1982.

19. Mehta R. A concise account of optics of magnetic fluids. In material aspects of ferrofluids, by R. Mehta (Taylor & Francis Group, 2023), p. 17.
20. Chen G, Jin Z, Chen J. A review: magneto-optical sensor based on magnetostrictive materials and magneto-optical material. Sens Actuators Rep. 2023;100152.
21. Ying Y, Si G-Y, Luan F-J, Xu K, Qi Y-W, Li H-N. Recent research progress of optical fiber sensors based on D-shaped structure. Opt Laser Technol. 2017;90:149–57.
22. Schukar V, Köppe E, Hofmann D, Westphal A, Sahre M, Gong X, et al. Magnetic field detection with an advanced FBG-based sensor device. Procedia Eng. 2016;168:1270–4.
23. Harry GM. Advanced LIGO: the next generation of gravitational wave detectors. Class Quantum Gravity. 2010;27(8): 084006.
24. Li X, Yu Q, Zhou X, Zhang Y, Lv R, Zhao Y. Magnetic sensing technology of fiber optic interferometer based on magnetic fluid: A Review. Meas. 2023;216: 112929.
25. Herrera-Piad LA, Haus JW, Jauregui-Vazquez D, Sierra-Hernandez JM, Estudillo-Ayala JM, Lopez-Dieguez Y, et al. Magnetic field sensing based on bi-tapered optical fibers using spectral phase analysis. Sensors. 2017;17(10):2393.
26. Cennamo N, Arcadio F, Marletta V, Baglio S, Zeni L, Andò B. A magnetic field sensor based on SPR-POF platforms and ferrofluids. IEEE Trans Instrum Meas. 2021;70:1–10.
27. Zhu L, Lin Q, Yao K, Zhao N, Yang P, Jiang Z. Fiber vector magnetometer based on balloon-like fiber structure and magnetic fluid. IEEE Trans Instrum Meas. 2021;70:1–9.
28. Zhou C, Ding L, Wang D, Kuang Y, Jiang D. Thinned fiber Bragg grating magnetic field sensor with magnetic fluid. SPIE. 2011;8034.
29. Rahul K. Magnetic field sensing using tapered small-core optical fibre surrounded by different concentrations of magnetic fluid. Sensors. 2022;22(21):8536.
30. Chen Y, Hu Y, Cheng H, Yan F, Lin Q, Chen Y, et al. Side-polished single-mode-multimode-single-mode fiber structure for the vector magnetic field sensing. J Light Technol. 2020;38(20):5837–43.
31. Ying ZNM, Dong X, Shum PP, Hu DJJ, Su H, Lew WS, et al. Magnetic field sensor based on magnetic-fluid-coated long-period fiber grating. J Opt. 2015;17(6): 065402.
32. Miao Y, Ma X, Lin J, Song B, Lin W, Zhang H, et al. Intensity-based magnetic field measurement employing tilted long-period fiber gratings. J Opt. 2015;17(10): 105609.
33. Cai S, Sergeev M, Petrov A, Varzhel S, Sheng C, Li L. Highly sensitive vector magnetic field sensors based on fiber Mach-Zehnder interferometers. Opt Commu. 2022;524: 128725.
34. Chen F, Jiang Y, Gao H, Jiang L. A high-finesse fiber optic fabry–perot interferometer based magnetic-field sensor. Opt Lasers Eng. 2015;71:62–5.
35. Lv R, Zhao Y, Wang D, Wang Q. Magnetic fluid-filled optical fiber Fabry–pérot sensor for magnetic field measurement. IEEE Photonics Technol Lett. 2014;26(3):217–9.
36. Peng J, Zhang S, Jia S, Kang X, Yu H, Yang S, et al. A highly sensitive magnetic field sensor based on FBG and magnetostrictive composite with oriented magnetic domains. Meas. 2022;189: 110667.
37. Han J, Hu H, Wang H, Zhang B, Song X, Ding Z, et al. Temperature-compensated magnetostrictive current sensor based on the configuration of dual fiber Bragg gratings. J Light Technol. 2017;35(22):4910–5.
38. Xu R, Niu G, Xue Y, Ke C, Deng H, Deng S, et al. An all-optical vector magnetic field sensor based on magnetic fluid and side-polished hollow-core optical fiber. IEEE Sens J. 2021;21(19):21410–6.
39. Chen Y, Sun W, Zhang Y, Liu G, Luo Y, Dong J, et al. Magnetic nanoparticles functionalized few-mode-fiber-based plasmonic vector magnetometer. Nanomaterials. 2019;9(5):785.
40. Zhang Y, Pu S, Li Y, Hao Z, Li D, Yan S, et al. Magnetic field and temperature dual-parameter sensor based on nonadiabatic tapered microfiber cascaded with FBG. IEEE Access. 2022;10:15478–86.
41. Xu J, Huang K, Zheng J, Li J, Pei L, You H, et al. Sensitivity enhanced magnetic field sensor based on hollow core fiber Fabry-Perot interferometer and vernier effect. IEEE Photon J. 2022;14(4):1–5.

42. Wang D, Yi Z, Ma G, Dai B, Yang J, Zhang J, et al. Two-channel photonic crystal fiber based on surface plasmon resonance for magnetic field and temperature dual-parameter sensing. Phys Chem Chem Phys. 2022;24(35):21233–41.
43. Homa D, Kaur G, Pickrell G, Scott B, Hill C. Electronic and magnetic fibers. Mater Lett. 2014;133:135–8.
44. Tang G, Sun M, Liu W, Qian Q, Qian G, Huang K, et al. Bismuth core fibers. Mater Lett. 2017;186:308–10.
45. Zachary H, Homa D, Ma L, Dong B, Martin E, Pickrell G, et al. Development of a multimaterial optical fiber for fully distributed magnetic sensing applications. IEEE Sens Lett. 2022;6(1):1–4.
46. Maayani S, Foy C, Englund D, Fink Y. Distributed quantum fiber magnetometry. Laser Photonics Rev. 2019;13(7):1900075.
47. Bucholtz F, Villarruel CA, Kirkendall CK, Dagenais DM, McVicker JA, Davis AR, et al. Fibre optic magnetometer system for undersea applications. Electron Lett. 1993;29:1032–3.
48. Bucholtz F, Wang G, Kirkendall CK, Villaruel CA, Davis AR, Dagenais DM, et al. Undersea performance of eight-element fibre optic magnetometer array. Electron Lett. 1994;30:1974–5.
49. Ou Y, Chen J, Chen W, Cheng C, Zhu Y, Xiao W, et al. A quasi-distributed fiber magnetic field sensor based on frequency-shifted interferometry fiber cavity ringdown technique. Opt Laser Technol. 2022;146: 107607.
50. Bai X, Hu M, Gang T, Rong Q. Simultaneous acoustic and magnetic measurement using cascaded fibre Bragg grating. Opt Fiber Technol. 2018;45:376–82.
51. Ma R, Zhang W, Huang W, Li F. DFB fiber laser sensor for simultaneous measurement of acoustic and magnetic fields. Chin Opt Lett. 2018;16: 060602.
52. Wang Z, Zhang W, Huang W, Liu P, Li F. A fiber optic accelerometer–magnetometer. J Light Technol. 2017;35(9):1732–7.

Chapter 5
Thermally Drawn Optoelectronic and Optofluidic Polymer Fibres for Brain Applications

Marcello Meneghetti, Kunyang Sui, and Christos Markos

Abstract In the last two decades, optical techniques for modulating and monitoring neural activity have changed the landscape of neuroscience. This has created a research drive towards the development of soft, biocompatible optical fibres able to deliver and collect light in deep brain regions over long periods of time with minimal inflammation. Thanks to their low Young's modulus and high versatility in terms of opto-mechanical properties, polymer-based fibres are an ideal platform to base this effort on. Furthermore, the interest in combining optical techniques with viral injections, pharmacology and electrophysiology while minimizing the number of surgeries has led to a strong interest in integrating different structures and functions in monolithic fibre-based devices. In this chapter, we will start by briefly introducing the most common functions that can be performed in the brain by elements integrated within polymer optical fibres. Afterwards, we will introduce the most commonly used polymers in neuroscience applications and describe all the steps required to fabricate optoelectronic and optofluidic fibres by thermal drawing process which is the most common technique for optical fibres fabrication. This will be followed by a discussion on the optimization of two parameters that are crucial for all multifunctional fibres targeted towards brain applications: the overall stiffness of the fibre, and the volume of tissue to which light can be delivered and from which light can be collected. Finally, we will introduce a recent application of the described technology to an emerging optical technique in the field of neuroscience: *infrared neuromodulation*.

M. Meneghetti · K. Sui · C. Markos (✉)
DTU Electro, Ørsteds Plads 343, 2800 Kgs., Lyngby, Denmark
e-mail: chmar@dtu.dk

M. Meneghetti
e-mail: mamen@dtu.dk

K. Sui
e-mail: kunsu@dtu.dk

L. Wei (ed.), *Advanced Optical and Optoelectronic Fibers*, Advances in Optics and Optoelectronics, https://doi.org/10.1007/978-981-97-6218-7_5

1 Introduction

Among the objectives of current research in neuroscience, one of the most crucial is the understanding of brain function and circuitry at the neuronal level, with the goal of gaining new insights on how the brain works towards developing new therapies for neurological and neurodegenerative diseases. In this framework, there is a strong necessity of tools able to modulate and record brain activity down to the single unit scale. Since the discovery of optogenetics [6, 48] and infrared neuromodulation [45] in the early 2000s, optical neuromodulation techniques have attracted significant attention in the field of neuroscience as a way to control neuronal activity with unsurpassed spatial and temporal resolution. Furthermore, aside from modulation, light provides a powerful alternative to electrophysiology to monitor the functions of neurons at the individual or population level [39]. These technological developments created a new requirement for implantable devices able to deliver and collect light to and from the brain. The simplest and most widespread implants for optical neuromodulation are silica fibres, similar to the ones commonly used in the field of telecommunications [43]. However, despite their excellent light transmission properties, broad availability and compatibility with most light sources, standard silica fibres have some considerable limitations when used as brain implants. The significant mechanical mismatch between silica and brain tissue leads to a strong foreign body response and inflammation when fibres are implanted for long periods of time, and the brittle and fragile nature of these fibres can cause them to fracture during the implantation procedure or during the movement of behaving animals, causing loss of functionality and potential release of sharp fragments in the tissue [29, 42]. Furthermore, the efficiency of light collection and delivery is limited by the numerical aperture and geometry of the fibres [36, 42]. Finally, standard optical fibres are limited to light-based applications, and the combination of optical techniques with other crucial tools in neuroscience, such as drug delivery and electrophysiology, requires complex multi-step surgical procedures [32]. Overcoming these limitations of standard silica optical fibres has been the focus of a large amount of research, and several different approaches have been exploited by groups around the world to develop new generations of neural interfaces with higher biocompatibility, better performance and integrated multiple functionalities.

Concerning the increase in biocompatibility, one of the most pursued strategies is the use of materials with lower Young's moduli compared to silica glass. Polymer optical fibres (POFs), in particular, have been proposed as an alternative to silica due to a number of advantages of this class of materials, such as low Young's modulus, low cost, and nontoxicity [44]. Other strategies include the use of tapered fibres to reduce the tissue damage during implantation, the use of biodegradable fibres to avoid the need to harmful explantation surgeries and a reduction in the footprint of devices to reduce the overall bending stiffness and invasiveness [32, 36, 38]. The efficiency of light delivery and collection can be optimized by changing the cross-section of the fibre, using tapered fibres or engineering the numerical aperture of the optical fibre [33, 36, 42]. Finally, the addition of functionalities to the fibres can be addressed by the integration of several optical, fluidic and electronic components within the fibre structure, either at the fabrication stage or through post-processing of the devices. These include, but are not limited to, fibre Bragg gratings (FBGs), Fabry–Perot interferometers, plasmonic sensors, engineered optical windows, electrodes and microfluidic channels [21–23, 25, 26, 28, 40, 42, 49].

In this chapter, in particular, we will focus on optoelectronic and optofluidic multifunctional implantable devices based on thermally-drawn POFs developed for brain applications, with a particular focus on functionalities enabling optical neuromodulation, electrophysiology, fluorescence photometry and delivery of drugs and viral vectors. The fabrication method for the fibres will be described step-by-step, and the optimization of the fibres' cross-sectional geometries, dimensions and numerical aperture towards highly biocompatible and efficient devices will be discussed in detail. Finally, we will introduce the open research field of soft implants for infrared neuromodulation.

2 Functionalities Integrated in Optical Fibres for Brain Applications

With the emergence of optical techniques for neuroscience, a lot of research focus has been devoted to the development of implantable optical devices for neuroscience. Optical fibres have attracted particular interest due to their flexibility, high transparency and scalable fabrication. Aside from the delivery of light to brain tissue for neuromodulation and phototherapy, optical fibres have been used for several brain-related applications, such as fluorescence photometry and sensing of temperature, oxygen and intracranial pressure [28]. The addition of functional elements to standard optical fibres, like electrodes and microfluidic channels, allowed the integration of non-optical applications such as pH sensing, electrophysiology and fluid delivery [3, 7, 18]. In this chapter, we will focus mainly on optoelectronic and optofluidic fibres targeted at the four applications listed below.

Optical neuromodulation has been the main driver of the research towards the use of optical fibres for the brain. The most common optical neuromodulation technique is optogenetics, consisting in the genetic manipulation of neurons to induce the expression of light-sensitive channel proteins (opsins) that can be activated or inhibited by shining visible light on them. Optogenetics rapidly became a widely used and extremely powerful neuromodulation technique, thanks to its high temporal resolution and the possibility to target specific neuronal subsets with the viral vectors used to achieve opsin expression. Infrared neuromodulation, on the other side, has recently gained traction as an alternative to optogenetics that does not require opsins. By using infrared light in spectral regions exhibiting high absorption from biological tissues, this technique relies on thermally induced effects resulting in the activation or inhibition of neurons. Though its biophysical mechanisms are not entirely understood, and despite the loss of cell-type specificity typical of optogenetics, infrared neuromodulation is emerging as a promising technique due to its enhanced spatial resolution and lack of requirement for genetic manipulation.

Electrophysiology is the most common technique used to record brain activity at the neuronal level. It relies on direct recording of the electrical potentials generated by neuronal membranes during activity and synaptic communication. Being a direct measurement, electrophysiology can be used to record neuronal activity in real time with excellent signal-to-noise ratio, with spatial resolutions ranging from large neuronal populations (local field potentials) to single neurons or membrane channels (action potentials).

Fluorescence photometry is emerging as a light-based alternative to electrophysiology for recording neuronal activity. The use of light instead of direct electrical measurements removes the requirement for direct contact between the probe and the neuronal membranes, allows for outstanding spatial resolution in recording and enables simultaneous recording in different regions. Similarly to optogenetics, fluorescence photometry relies on reporters able to convert neuronal activity to light emission, a fact that enables cell-specificity through targeted manipulation. However, being an intrinsically indirect measurement, fibre photometry requires a complex calibration to obtain quantitative results [39].

Delivery of drugs and viral vectors is crucial for a range of applications in neuroscience. Targeted delivery of drugs to specific brain regions can be a tool both in the understanding of brain circuitry and in the development of new pharmaceutical compounds, and can be combined with optical and electrical techniques to verify in real time the effect of drugs at a local level. Furthermore, the delivery of viral vectors with high precision to target brain regions is crucial for both optogenetics and fluorescence photometry, since they rely on genetic manipulation as their base mechanism.

3 Polymers for Brain-Compatible Multifunctional POFs

When designing a step-index optical fibre, the choice of the core and cladding materials is crucial. This selection must be done by keeping into account the transmission windows of the core material, which should overlap with the wavelengths needed for the specific application, the refractive indices of both materials, which should satisfy Snell's law ($n_{core} > n_{cladding}$) in order to achieve light guidance, and the thermal compatibility between materials (similar glass transition temperatures, melting temperatures and viscosity profiles) which is crucial to be able to successfully co-process the materials by thermal fibre drawing. When the intended application for the optical fibre is implantation in soft biological tissues, such as the brain, other properties must also be taken into account, most notably the nontoxicity, biocompatibility and Young's modulus of the material. This last parameter has been proven to be of particular importance due to its role in determining the final bending stiffness of the implant, which is crucial to limit inflammation during chronic experiments. Due to their low Young's moduli when compared with other optical materials, in recent years polymers have been often used as the materials of choice for the step-index waveguide when developing multifunctional optical fibres for brain implants. A large variety of thermoplastic polymers have been used to develop soft fibres with different properties and functionalities. The refractive indices, melting temperatures, glass transition temperatures and Young's moduli of the polymers mentioned in this section are reported in Table 1.

In 2015, Canales et al. developed the first multimaterial, multifunctional implantable devices able to simultaneously deliver light, record electrical signals and inject viral vectors for optogenetics based on a thermally drawn POF with polycarbonate (PC) core and cyclic-olefin-copolymer (COC) cladding. The waveguide had losses in the visible compatible with light delivery in the brain of rodents (<3 dB/cm) and showed considerably lower bending stiffness when compared to silica, resulting in reduced glial response from the brain tissue over 3 months of implantation in mice [7, 32]. Similarly, PC has been used as core material by Jiang et al. to develop multifunctional POF-based neural interfaces, here with polyvinylidene difluoride (PVDF) as a cladding material [20]. COCs have also been used as core material for optoelectronic brain fibres, both with polymethyl methacrylate (PMMA) cladding and with a different COC as cladding [11, 41]. Recently, we have demonstrated for the first time the use of fluorinated ethylene propylene copolymer (FEP) as a cladding material for multifunctional optical fibres with PC and polysulfone (PSU) polymers [42]. This material is of particular interest because of its extremely low Young's modulus, 4–5 times lower than the one of common optical polymers, and its low refractive index allowing for the fabrication of ultra-high numerical aperture fibres with enhanced illumination and light collection properties. Interestingly, PSU-FEP fibres also exhibit an extended transmission range in the near infrared, and have been used to develop the first polymer optical fibre able to achieve infrared neuromodulation in the high-water-absorption spectral region close to 2 μm [25, 26]. Measured attenuation spectra for the core materials mentioned in this section are presented in Fig. 1.

Table 1 Properties of thermally drawable polymers used for the development of multifunctional neural interfaces, compared with the ones of standard silica (obtained from: [5, 7, 30, 42, 46])

Material	Refractive index	Glass transition temperature (°C)	Melting temperature (°C)	Young's modulus (GPa)
PC	1.577–1.587	137–154	255–267	2.39–2.6
COC	1.51–1.54	62–177	–	2.4–3.2
PVDF	1.42–1.49	−29–−57	185–200	2.5–2.7
PMMA	1.484–1.492	82–105	105–160	5
FEP	1.338–1.344	80	250–280	0.48
PSU	1.633	185–190	–	2.69
PEI	1.630–1.687	209–249	229	3
PPSU	1.67	220	–	2.3
Silica	1.45–1.47	1207	1734	70

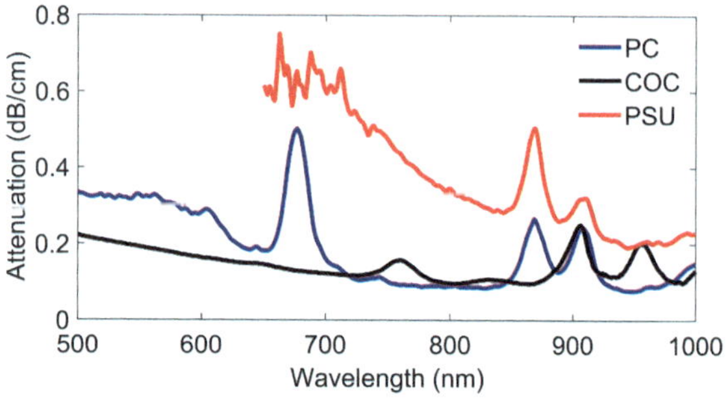

Fig. 1 Measured attenuation spectra for PC, COC and PSU (adapted from [41])

Aside from the waveguide itself, polymeric materials such as PSU, polyetherimide (PEI) and polyphenylsulfone (PPSU) have been used in multifunctional fibre-based neural devices to act as surrounding for electrodes, or as sacrificial layers to be etched during post-processing to reduce the size of the fibres [32, 41].

4 Thermal Drawing of Multimaterial and Multifunctional POFs

Once the polymers constituting the fibres have been selected, the next step in the development of POF-based brain implants is the fabrication of the optical fibre itself. In the case of POF for neural applications, the most common fabrication technique is thermal drawing process, due to its relative simplicity, versatility and high scalability. In this section, we will describe the steps required to fabricate a multifunctional optical fibre by thermal drawing process, from the fabrication of the initial preform to the integration of multifunctional elements within the fibre.

- **Preform fabrication**

The first step in thermal fibre drawing is the fabrication of a macroscopic template of the fibre's geometry, known as preform. Here, we will discuss the most common techniques to obtain a multimaterial step-index preform starting from raw materials in different forms.

Rod-in-tube technique is the simplest and most commonly used approach for building multimaterial step-index preform. Here, a rod composed of the core material is fitted in a hole drilled in a larger cylinder of cladding material, directly creating a template of the core-clad structure of the fibre (Fig. 2a). Additional holes might be drilled in as the cladding material for integrating microfluidic channels or electrodes, as will be discussed later in this chapter.

Sheet rolling uses thin polymer sheets rather than tubes as a starting form for the cladding material. The sheets are tightly rolled around the core rod, until a cladding of the desired thickness is formed (Fig. 2b). Following the rolling step, the preform needs to be consolidated under vacuum at temperatures above the cladding polymer's

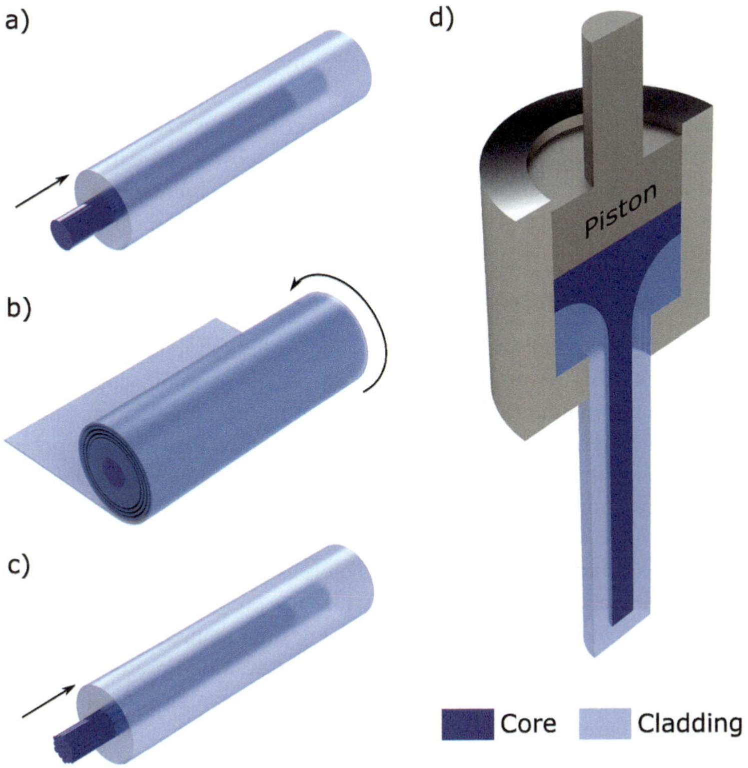

Fig. 2 Fabrication techniques for multimaterial preforms. **a** Rod-in -tube technique; **b** Rolling of polymer sheet around a core rod; **c** Stack-and-draw technique; **d** Co-extrusion of core and cladding materials

glass transition temperature in order to form a gapless solid rod. This technique can be used in multiple steps, and combined with machining or etching of gaps to create complex cross-sectional geometry. Canales et al. used this technique in 2015 to develop the preform for the first multifunctional POF for brain applications [7].

Stack-and-draw is a common approach for the fabrication of microstructured, nanostructured and photonic crystal fibres [23, 24, 37]. The preform is built by stacking several canes and/or capillaries in a predetermined geometry, which are then inserted in an external jacket (Fig. 2c). Different materials can be used for different elements of the structure, and repeated drawing steps can be used to develop structured canes (e.g. waveguides, insulated electrodes) with different functionalities to be combined in a single final preform. Recently, Antonini et al. employed the stack-and draw technique to develop fMRI compatible multifunctional POFs with integrated optical, electrical and fluidic modalities [4]. Jiang et al. instead used the stack-and-draw approach to fabricate PC scaffolds used to direct the tip of multifunctional fibres to different brain locations [20].

Extrusion is a process employing temperature and pressure to generate preforms and canes with a fixed cross-sectional profile, by pushing a soft material through a die. This technique has been commonly used to develop preforms with complex geometries for microstructured and suspended-core fibres [15, 47]. In the case of multimaterial fibres, of particular interest is the process of co-extrusion, in which a stack of disks of different materials can be extruded simultaneously to create a concentric structure, such as a step-index core-clad preform (Fig. 2d), with extremely high-quality interface between the core and cladding materials [1].

- **Thermal drawing process**

Thermal drawing is the most common process for the fabrication of optical fibres, using pulling under high temperature to scale the preform to microscopic temperature. Due to its high throughput (kilometers of optical fibres can be drawn from a single preform) and the high degree of control on the fibre final diameter and geometry, it is used to produce the millions of kilometers of silica optical fibres that constitute the global telecommunications network.

At the start of the process, the preform is loaded vertically into a furnace, in which the temperature is increased until all the materials are softened (necking). Finally, the fibre is pulled at a constant speed v_f while the preform is fed into the furnace with speed v_p to scale down the material to the desired diameter (Fig. 3), monitored by a gauge during the process. The ratio between the initial preform diameter and the final fibre diameter (draw-down ratio, *DD*) can be expressed as a function of the fibre drawing and preform feeding speeds as

$$DD = \sqrt{\frac{v_f}{v_p}}$$

- **Integration of electrodes and microfluidic channels**

The development of optoelectronic and optofluidic POFs requires the direct integration within the fibre of electrodes and microfluidic channels, respectively, either during the thermal drawing process or using a post-processing approach on the fabricated fibre. The integration of microfluidic channels can be regarded as straightforward, and usually results from the introduction of holes or hollow channels in the preform that will be scaled down to microscopic size during the drawing (Fig. 4a) [7, 42]. The integration of electrodes, instead, requires further considerations due to the broad range of conductive materials that can be employed, and to the difference in their thermo-mechanical properties with respect to the ones of the polymers constituting the step-index fibre. Among the different possible approaches to integrate metallic electrodes into polymer fibres, mainly three have been used for neural applications.

Co-drawing of polymers and conductive elements consists in the integration of macroscopic electrodes at the preform stage that will be scaled down together with

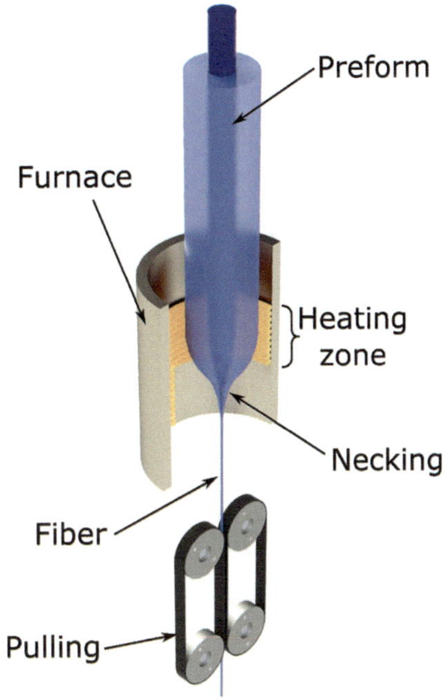

Fig. 3 Schematic of thermal drawing process

the rest of the fibre geometry (Fig. 4a), and thus requires to select conductive materials that are thermally compatible with the polymers constituting the fibre. This method is therefore limited to low melting point metals and alloy, or to conductive polymers. As an example of polymer-compatible metals, tin and indium have been used for the integration of electrodes into polymer fibres by co-drawing them with PSU or within a PC-COC preform, respectively [4, 7]. As an alloy, BiSn has been reported to be compatible for thermal drawing with PC, PMMA, COC and PVDF [17, 20]. Despite their significantly lower conductivity with respect to metals and alloys, conductive polymers such as conductive polyethylene have also been successfully used for developing optoelectronic neural interfaces [7]. Interestingly, the addition of graphite (5% w/w) to CPE resulted in a fourfold reduction of the material's sheet resistivity, allowing for a reduction of the electrode's size [32].

Hot–cold drawing is a technique to integrate into POFs metals having melting temperatures significantly higher than the ones of polymers. Through holes are machined into the preform, and metallic microwires are directly fed in the holes during the drawing process (Fig. 4b) [27]. Unlike co-drawing, in this case the metal wires, having melting temperatures up to thousands of °C higher than the ones of polymers, maintain their cross-section during the process and are integrated into the fibre thanks to the holes collapsing around them. This process has been used to integrate highly conducting metals, such as copper and tungsten (melting temperatures of 1080 °C and 3410 °C, respectively [12]), into optoelectronic brain interfaces based

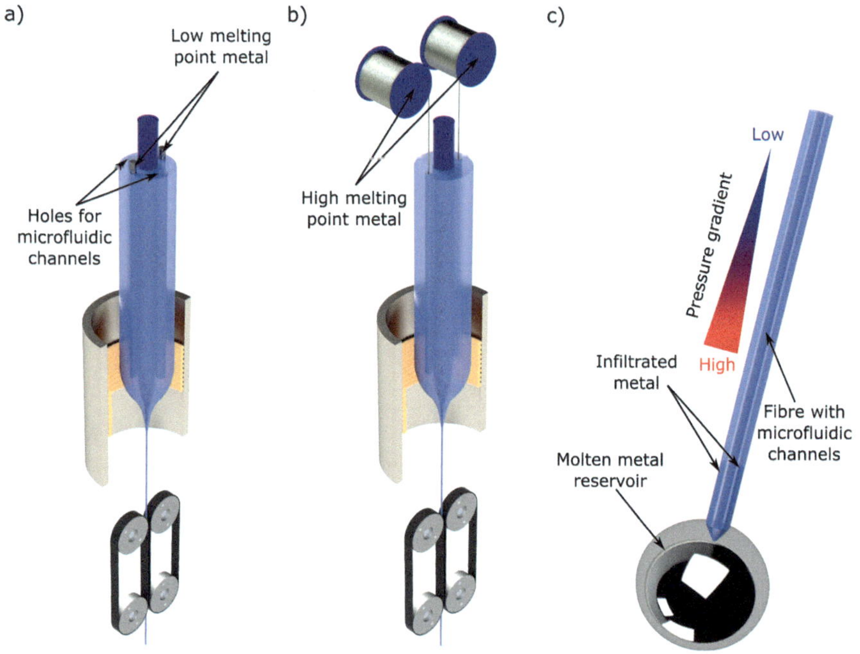

Fig. 4 Schematics of microfluidic channels and electrodes integration methods. **a** Integration of low melting point metals and microfluidic channels by co-drawing; **b** Integration of high melting point metals by hot–cold drawing; **c** integration of electrodes by metal infiltration

on low drawing temperature polymers such as PC, PVDF, PMMA, PSU and COC [4, 13, 41].

Infiltration of metals, finally, allows to integrate metals with melting temperatures lower than the ones of polymers, and which tend to oxidize during the prolonged heating required by the thermal fibre drawing process. To avoid this issue, fibres are drawn with hollow channels designed to accommodate the electrodes, and only after the drawing the metallic elements are integrated into the multifunctional device by using pressure to inject molten metal into the fibres (Fig. 4c). We used this procedure to integrate low melting point indium electrode within a POF composed of high-performance polymers with high drawing temperatures (PSU and FEP) [26].

5 Optimization of the Fibre's Properties Towards Brain Applications

In order to be suitable for the complex task of interrogating neurons and brain circuitry, optoelectronic and optofluidic fibres require a careful optimization of several optical, electrical and microfluidic properties, depending on the specific functionalities integrated in the fibres and intended uses. In this section, we will focus in particular on two aspects that are common to all multifunctional fibre-based implants: the optimization of the fibre's stiffness to achieve minimum inflammation in the brain tissue while maintaining the possibility of successful implantation, and the optimization of the volume and distribution of the light that can be delivered to or collected from the brain tissue through the optical fibre's core.

- **Optimization of flexibility through geometry**

When developing implantable devices, one of the crucial parameters to consider is their flexibility, often described in terms of their bending stiffness (the force F necessary to achieve a deflection d of the implant). The higher the bending stiffness, the larger the damage and inflammation caused by the implant in chronic settings due to the constant relative micromotion of brain and implant. The bending stiffness can be expressed as a function of the implant's composite Young's modulus E, its momentum of inertia I and its length L [10]:

$$\frac{F}{d} = \frac{48E \bullet I}{L^3}$$

Therefore, aside from the individual Young's modulus of the materials constituting the fibre, it is crucial to consider the fibre's cross-sectional geometry and implantable length to optimize its biocompatibility and ensure its long-term stability in vivo. For simple step-index optical fibres, once the materials and core-clad ratio are fixed, the relationship between geometry and total bending stiffness is quite straightforward: the bending stiffness increases for larger fibre diameters, and decreases for longer implants. When functional elements such as electrodes and microfluidic channels are included in the structure, however, also the diameter and relative distances of such structures need to be taken into account. Furthermore, these elements will break the circular symmetry of the fibre, introducing a dependence of the bending stiffness on the bending direction.

Recently, we used a finite elements methods numerical model based on Von Mises stress distribution to evaluate the influence of these parameters on the fibre's overall bending stiffness. For all simulations, a step-index fibre with a core-clad ratio of 2:5 and two circular functional elements (electrodes or microfluidic channels) embedded in the cladding in diametrically opposed positions with respect to the core has been used as model geometry [41]. Our model shows that the impact of the functional element on the overall bending stiffness of the implant mainly concerns forces directed along the axis connecting the centers of the functional elements, while the bending stiffness in the orthogonal direction is comparatively unaffected. The inclusion of electrodes, especially ones composed of stiff materials such as tungsten (Young's modulus of 360 GPa [9]) shows a strong effect on the bending stiffness, which increases both when increasing the elements size and when increasing the distance between in them (Fig. 5a, 5b). Inclusion of microfluidic channels, on the other side, causes an overall reduction in the bending stiffness, which becomes more pronounced when increasing their size or relative distance (Fig. 5c, d).

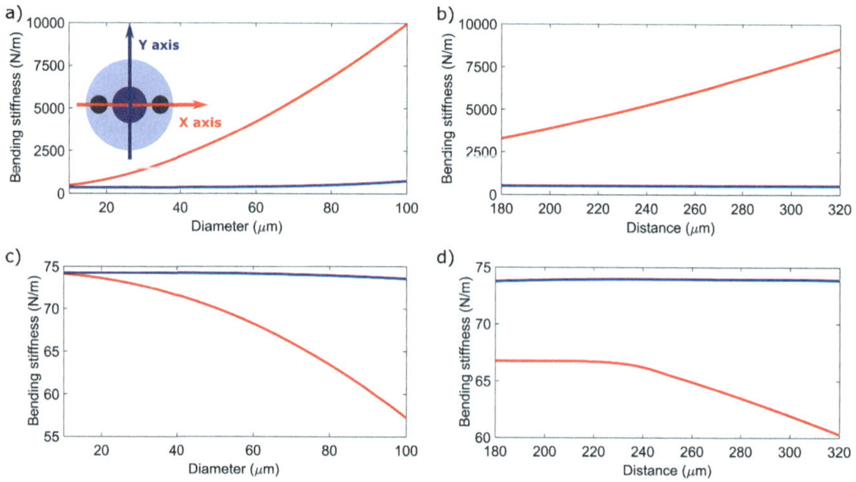

Fig. 5 Impact of diameter and relative distance of microfluidic channels and tungsten wires on the bending stiffness of an implant (adapted from [41]). **a**, **b** Bending stiffness plotted against the diameter and distance of tungsten microwires, respectively. The model is a fibre with COC core and PSU cladding, a total diameter of 400 μm, a length of 3 mm; **c**, **d** Bending stiffness plotted against the diameter and distance of microfluidic channels, respectively. The model is a fibre with PSU core and FEP cladding, a total diameter of 400 μm, a length of 3 mm

While reducing the overall stiffness of the fibre has a positive effect on the brain tissue once the device is implanted, it is also important to consider its effect on the implantation procedure itself. If an implant is too soft, this might cause it to bend during the penetration of the tissue (buckling), causing either an inaccurate targeting of the brain regions or a complete failure of the implantation itself. This effect is caused by the load along the fibre's axis experienced by the implant during implantation, which is known to reach its maximum value, P_i, of about 1 mN when the fibre penetrates the dura layer [42]. The risk of buckling failure can be described by using a critical load factor parameter, defined as the ratio between the critical load required to bend the fibre (P_c) and the maximum load during implantation.

$$Critical\ load\ factor = \frac{P_c}{P_i}$$

A 100% success rate of implantation against buckling failure corresponds to a critical load factor of 3.5. Since, as stated before, the overall implant stiffness decreases with increasing length and increases with increasing diameter, it's possible to define a critical length below which, for a fixed diameter, the implantation will be successful. Similarly, once the length is fixed, a critical diameter above which the risk of buckling is non-existent can be calculated. At a fixed length, the integration of microfluidic channels will increase the critical diameter, while the introduction of electrodes stiffer than the surrounding polymer will decrease it. Thus, further miniaturization of the implants will require additional strategies to avoid buckling failure, such as the use of stiff guides during implantation or biodegradable coatings to temporarily increase the diameter above the critical one.

- **Optimization of illumination profile and collection efficiency**

For optical techniques in neuroscience, such as optical neuromodulation and fluorescence photometry, the volume of tissue that can be illuminated or recorded is a crucial parameter. For neuromodulation, depending on the intended application, the ability to illuminate larger volume can be crucial to verify brain circuitry involving larger brain regions. Furthermore, the intensity profile of the light within the brain tissue can increase or reduce the local fluence in the area surrounding the fibre tip, thus affecting the maximum total power that can be delivered through the fibre without surpassing the threshold for light-induced tissue damage [14]. Similarly, for fluorescence photometry a large illumination and collection volume allows the simultaneous interrogation of a larger number of neurons, significantly increasing the advantage of this technique as an alternative to electrophysiology for monitoring neuronal activity.

A possible strategy to increase the illumination volume and collection efficiency of implantable optical fibres is to use tapered fibres [36]. Thanks to the modal light propagation properties of this class of fibres, it is possible to use them to simultaneously illuminate the whole tissue region surrounding the taper, achieving uniform delivery of light along large depths of brain rather than just in a limited region of the fibre's tip. Similarly, the collection of signal from fluorescent reporters is uniformly spread along the fibre. The functionality of tapered fibres can also be increased by designing a non-transparent coating with optical windows, that can be individually addressed by changing the light coupling conditions, leading to a selective optogenetic stimulation along the taper's length [35]. However, this approach has been only applied to glass optical fibres.

For standard flat-cleaved POFs, the main strategies to increase the illumination and collection efficiency are related to the optimization of the fibre's core size and numerical aperture. Both these parameters, when increased, lead to a larger lateral spread of the delivered light within the brain tissue, an effect that is stronger in the proximity of the fibre's tip. This has the twofold effect of increasing the number of illuminated neurons and reducing the risk of thermal damage [19]. Similarly, the collection efficiency in fibre photometry is positively correlated to both numerical aperture and core diameter. In this section, the focus will be on increasing the volume of tissue that can be targeted, all of the above considerations on core diameter and numerical aperture can easily be reversed for applications that require instead a high spatial resolution and thus a minimization of the affected volume.

Due to the correlation between the fibre's overall diameter and its bending stiffness discussed in the previous section, increasing the illumination and collection efficiency of POFs for brain applications through increasing the core's diameter can lead to detrimental effects in terms of inflammatory foreign body response in long-term implantations. Thanks to the wide variety of polymeric materials that can be used for developing multifunctional optical fibres, on the other side, it's possible to tune their numerical apertures over a broad range.

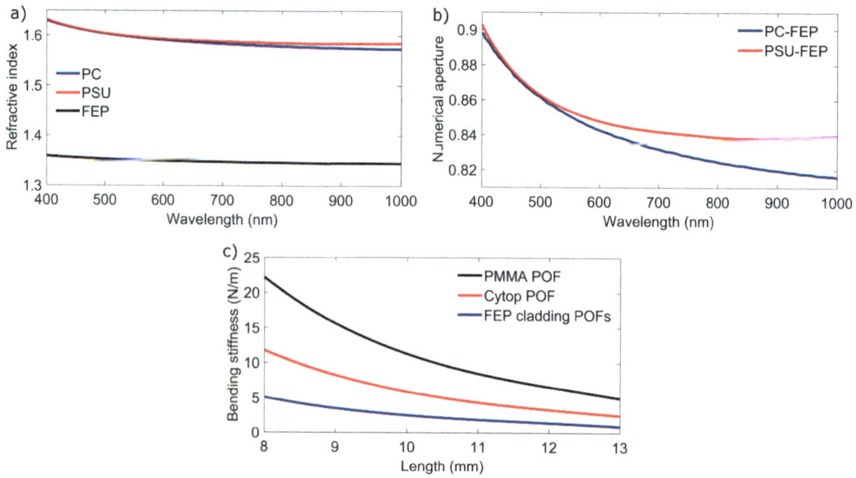

Fig. 6 Optical and mechanical properties of ultra-high numerical aperture optical fibres with FEP cladding (adapted from [42]). **a** Measured refractive indices of the core (PC and PSU) and cladding (FEP) polymers used for developing the fibres; **b** calculated numerical apertures for the two types of fibre; **c** bending stiffness, plotted against implant length, for two commercial POFs (PMMA and Cytop) and for the developed POFs with FEP cladding

In particular, recently we demonstrated for the first time the use of FEP as cladding material in a step index POF [42]. Having one of the smallest refractive indices among thermally drawable polymers (close to the one of water), the use of this material significantly increases the limit of numerical aperture reachable by flat-cleaved, step index POFs. As core materials, we used PC and PSU, both having a refractive index difference with FEP of more than 0.2, leading to an ultra-high numerical aperture larger than 0.8 (Fig. 6a, b).

The increase in illumination efficiency with respect to a standard silica fibre used for optogenetics (numerical aperture of 0.39) with similar core diameter was experimentally verified by measuring the 2-D intensity profile at the fibre output. This experiment was performed both in a standard brain phantom (0.6% w/w agarose in water) commonly used to simulate brain properties [2] and in perfused brain slices. The results of these experiments show that the FEP-clad fibres exhibit an overall enlarged illumination area, with an increase with respect to silica of 17% at a wavelength of 470 nm and of 21% at a wavelength of 650 nm in brain phantom, corresponding to an increased efficiency in both optogenetics and fibre photometry applications for large volumes. In the brain slices measurements, we were able to show a significant enhancement of the light's lateral spread, and thus a reduction of the fluence in the proximity of the fibre tip and of the risk of optical damage. Aside from allowing higher numerical apertures, the extremely low Young's modulus of FEP significantly decreased the overall bending stiffness of the POFs when compared with standard commercially available POFs, such as PMMA fibres and graded index Cytop fibres (Fig. 6c).

6 Towards Multifunctional POFs for Infrared Neuromodulation

Due to the fact that most common optical polymers have their transparency spectral windows in the visible and near infrared range, their applications to optical neuromodulation have almost exclusively been limited to optogenetics. However, infrared neuromodulation is attracting attention as a transgene-free alternative to optogenetics, and has been successfully applied in several contexts, which have been recently expanded to experimentation in the human brain [16, 31, 34]. This has evidenced the large technological gap between the broad variety of soft, biocompatible multifunctional devices that have been developed for optogenetic applications and the limited amount of implants dedicated to infrared neuromodulation. Thus, a large field of research is left open towards the identification of polymers and soft material able to transmit light in the spectral range of high tissue absorption (wavelengths > 1800 nm).

Recently, we have developed the first step-index POF able to deliver light at the wavelengths required for infrared neuromodulation, constituted of a PSU core surrounded by an FEP cladding [26]. This fibre has been used as the base to fabricate implantable devices with integrated electrodes to simultaneously elicit neuronal activity with infrared light and record it by electrophysiology (Fig. 7a). By coupling light from a broadband supercontinuum source into the core, we obtained an output spectrum overlapping with the main near infrared absorption peak of water (the main absorber in biological tissues) at ~1930 nm (Fig. 7b). Due to the high melting temperatures of both FEP and PSU, it was possible to integrate indium electrodes within the devices by using the metal infiltration method. Electrochemical impedance spectroscopy in phosphate-buffered saline solution revealed an average tip impedance of less than 25 kΩ at 1 kHz (Fig. 7c), significantly below the maximum threshold of 1 MΩ at 1 kHz recommended for electrophysiology [8]. The developed POF-based optoelectronic implantable devices allowed us to repeatably stimulate neuronal activity in the vicinity of the fibre's core with a 120 s ON / 120 s OFF stimulation protocol, and to record the elicited action potentials with high signal-to-noise ratio (Fig. 7d).

7 Conclusion

In summary, this chapter discussed the applications, fabrication and optimization of optoelectronic and optofluidic polymer optical fibres towards uses in neuroscience and neurotechnology. The main functionalities that have been integrated in these fibres, and their relevance to the investigation of brain function and circuitry, have been introduced. A list of common optical polymers used by different research groups across the world to realize these fibres has been provided, complete with their most

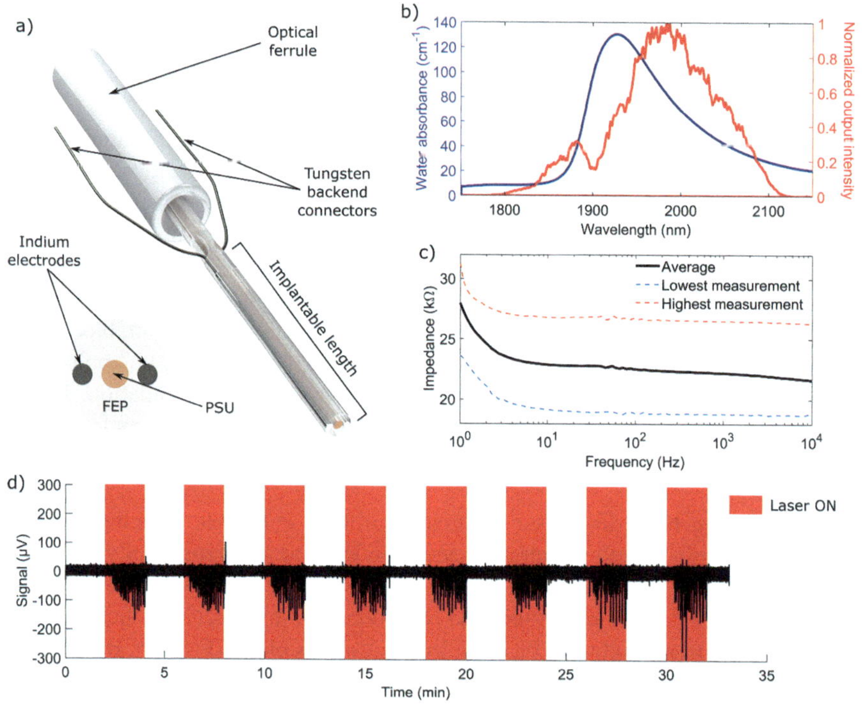

Fig. 7 Soft implantable implant for simultaneous infrared neuromodulation and electrophysiology (adapted from [25, 26]). **a** Schematic representation of the implant; **b** Overlap between the spectrum of the implant's optical output and the main near-infrared absorption peak of water; **c** Average, minimum and maximum tip impedance spectra measured from the indium electrodes (N = 6); **d** Neuronal activity elicited by infrared neuromodulation delivered through the implant and recorded by the integrated electrodes

relevant properties for the development of neural implants. The fabrication of multi-material and multifunctional optical fibres by thermal drawing process has been presented in detail through all its steps, starting from the preform fabrication down to the integration of functional elements. The optimization of the structural and optical parameters of the fibres to enhance the biocompatibility and reduce inflammation, ensure a successful implantation in the brain and increasing the affected tissue volume while avoiding phototoxicity have also been thoroughly discussed. Finally, a recently developed application of optoelectronic POFs for infrared neuromodulation has been presented.

References

1. Akrami P, Adamu AI, Woyessa G, et al. All-polymer multimaterial optical fiber fabrication for high temperature applications. Optical Materials Express. 2021;11:345–54.
2. Allison SW, Gillies GT. Fluorescence excitation and propagation through brain phantom gelatins: measurements and potential applications. Meas Sci Technol. 2010;21: 085802.
3. Anikeeva P, Andalman AS, Witten I, et al. Optetrode: a multichannel readout for optogenetic control in freely moving mice. Nat Neurosci. 2012;15:163–70.
4. Antonini M-J, Sahasrabudhe A, Tabet A, et al. Customizing MRI-compatible multifunctional neural interfaces through fiber drawing. Adv Func Mater. 2021;31:2104857.
5. Bansal NP, Doremus RH. Handbook of glass properties. Elsevier; 2013.
6. Boyden ES, Zhang F, Bamberg E, et al. Millisecond-timescale, genetically targeted optical control of neural activity. Nat Neurosci. 2005;8:1263–8.
7. Canales A, Jia X, Froriep UP, et al. Multifunctional fibers for simultaneous optical, electrical and chemical interrogation of neural circuits in vivo. Nat Biotechnol. 2015;33:277–84.
8. Canales A, Park S, Kilias A, Anikeeva P. Multifunctional fibers as tools for neuroscience and neuroengineering. Acc Chem Res. 2018;51:829–38.
9. Cui J, Ma L, Chen G, et al. Effect of twin boundaries on the strength of body-centered cubic tungsten nanowires. Mater Sci Eng A. 2023;862: 143826.
10. Dai Y, Xue Y, Zhang J. Vibration-based milling condition monitoring in robot-assisted spine surgery. IEEE/ASME Trans Mechatron. 2015;20:3028–39.
11. Davey CJ, Argyros A, Fleming SC, Solomon SG. Multimodal optogenetic neural interfacing device fabricated by scalable optical fiber drawing technique. Appl Opt. 2015;54:10068–72.
12. Davis JR. Metals handbook desk edition. (No Title) (1998)
13. Du M, Huang L, Zheng J, et al. Flexible fiber probe for efficient neural stimulation and detection. Adv Sci. 2020;7:2001410.
14. Dubois A, Chiang C-C, Smekens F, et al. Optical and thermal simulations for the design of optodes for minimally invasive optogenetics stimulation or photomodulation of deep and large cortical areas in non-human primate brain. J Neural Eng. 2018;15: 065004.
15. Ebendorff-Heidepriem H, Monro TM. Extrusion of complex preforms for microstructured optical fibers. Opt Express. 2007;15:15086–92.
16. Fekete Z, Horváth ÁC, Zátonyi A. Infrared neuromodulation: a neuroengineering perspective. J Neural Eng. 2020;17: 051003.
17. Guo Y, Werner CF, Canales A, et al. Polymer-fiber-coupled field-effect sensors for label-free deep brain recordings. PLoS ONE. 2020;15: e0228076.
18. Guo Y, Werner CF, Handa S, et al. Miniature multiplexed label-free pH probe in vivo. Biosens Bioelectron. 2021;174: 112870.
19. Gysbrechts B, Wang L, Trong NND, et al. Light distribution and thermal effects in the rat brain under optogenetic stimulation. J Biophotonics. 2016;9:576–85.
20. Jiang S, Patel DC, Kim J, et al. Spatially expandable fiber-based probes as a multifunctional deep brain interface. Nat Commun. 2020;11:6115.
21. Kalli K, Riziotis C, Posporis A, et al. Flat fibre and femtosecond laser technology as a novel photonic integration platform for optofluidic based biosensing devices and lab-on-chip applications: Current results and future perspectives. Sens Actuators, B Chem. 2015;209:1030–40.
22. Markos C, Stefani A, Nielsen K, et al. High-T g TOPAS microstructured polymer optical fiber for fiber Bragg grating strain sensing at 110 degrees. Opt Express. 2013;21:4758–65.
23. Markos C, Travers JC, Abdolvand A, et al. Hybrid photonic-crystal fiber. Rev Mod Phys. 2017;89: 045003.
24. Meneghetti M, Forestier X, Petersen CR, et al. Graded index chalcogenide fibers with nanostructured core. Adv Photonics Res. 2021;2:2000091.
25. Meneghetti M, Gudmundsen F, Jessen NS, et al. Mapping whole brain effects of infrared neural stimulation with positron emission tomography. Imaging Neurosci. 2023;1:1–17.

26. Meneghetti M, Kaur J, Sui K et al. Soft monolithic infrared neural interface for simultaneous neurostimulation and electrophysiology. Light Sci Appl. 2023b;12:127
27. Meneghetti M, Petersen CR, Hansen RE, et al. Thermally tunable dispersion modulation in a chalcogenide-based hybrid optical fiber. Opt Lett. 2021;46:2533–6.
28. Nazempour R, Zhang B, Ye Z et al. Emerging applications of optical fiber-based devices for brain research. Adv Fiber Materials. 2022;1–19
29. Nazempour R, Zhang Q, Fu R, Sheng X. Biocompatible and implantable optical fibers and waveguides for biomedicine. Materials. 2018;11:1283.
30. Ojovan MI, Tournier RF. On structural rearrangements near the glass transition temperature in amorphous silica. Materials. 2021;14:5235.
31. Pan L, Ping A, Schriver KE, et al. Infrared neural stimulation in human cerebral cortex. Brain Stimul. 2023;16:418–30.
32. Park S, Guo Y, Jia X, et al. One-step optogenetics with multifunctional flexible polymer fibers. Nat Neurosci. 2017;20:612–9.
33. Petrovic J, Lange F, Hohlfeld D. Fiber-based optrode with microstructured fiber tips for controlled light delivery in optogenetics. J Neural Eng. 2023;20: 036007.
34. Ping A, Pan L, Zhang J, et al. Targeted optical neural stimulation: a new era for personalized medicine. Neuroscientist. 2023;29:202–20.
35. Pisanello F, Sileo L, Oldenburg IA, et al. Multipoint-emitting optical fibers for spatially addressable in vivo optogenetics. Neuron. 2014;82:1245–54.
36. Pisano F, Pisanello M, Lee SJ, et al. Depth-resolved fiber photometry with a single tapered optical fiber implant. Nat Methods. 2019;16:1185–92.
37. Pysz D, Kujawa I, Stępień R, et al. Stack and draw fabrication of soft glass microstructured fiber optics. Bulletin Polish Acad Sci Technical Sci. 2014;62:667–82.
38. Sarabi MR, Jiang N, Ozturk E, et al. Biomedical optical fibers. Lab Chip. 2021;21:627–40.
39. Scanziani M, Häusser M. Electrophysiology in the age of light. Nature. 2009;461:930–9.
40. Spagnolo B, Balena A, Peixoto RT, et al. Tapered fibertrodes for optoelectrical neural interfacing in small brain volumes with reduced artefacts. Nat Mater. 2022;21:826–35.
41. Sui K, Meneghetti M, Berg RW, Markos C. Optoelectronic and mechanical properties of microstructured polymer optical fiber neural probes. Opt Express. 2023;31:21563–75.
42. Sui K, Meneghetti M, Kaur J, et al. Adaptive polymer fiber neural device for drug delivery and enlarged illumination angle for neuromodulation. J Neural Eng. 2022;19: 016035.
43. Vázquez-Guardado A, Yang Y, Bandodkar AJ, Rogers JA. Recent advances in neurotechnologies with broad potential for neuroscience research. Nat Neurosci. 2020;23:1522–36.
44. Wang Y, Huang Y, Bai H, et al. Biocompatible and biodegradable polymer optical fiber for biomedical application: a review. Biosensors. 2021;11:472.
45. Wells J, Kao C, Mariappan K, et al. Optical stimulation of neural tissue in vivo. Opt Lett. 2005;30:504–6.
46. Wypych G. Handbook of polymers. Elsevier; 2022.
47. Xue Z, Liu S, Zhao Z, et al. Infrared suspended-core fiber fabrication based on stacked chalcogenide glass extrusion. J Lightwave Technol. 2018;36:2416–21.
48. Zemelman BV, Lee GA, Ng M, Miesenböck G. Selective photostimulation of genetically chARGed neurons. Neuron. 2002;33:15–22.
49. Zheng D, Pisano F, Collard L, et al. Toward plasmonic neural probes: SERS detection of neurotransmitters through gold-nanoislands-decorated tapered optical fibers with Sub-10 nm Gaps. Adv Mater. 2023;35:2200902.

Chapter 6
Advanced Optical and Optoelectronic Fibers for Bio-integrated Systems

Zhiyuan Meng and Mengxiao Chen

Abstract This chapter highlights the potential of optical and optoelectronic fibers-based devices, circuits, and systems in bio-integrated applications. With advancements in material design and mechanics, fibers are being developed in stretchable, flexible, and biocompatible formats, allowing for customized integration with biological systems. We will discuss materials selection and three fabrication methods, and highlight specific applications in detecting biophysical signals, biosensors, human–machine interfaces, and photo-biomodulation therapy, among others. Finally, we will address remaining challenges in bio-integrated systems and provide practical insights.

Keywords Optical and optoelectronic fibers · Bio-integrated system · Biocompatible · Physiological signals · Human–machine interfaces · Photobiomodulation therapy

1 Introduction

The advent of contemporary optical fiber technology has revolutionized various fields, from data communication and healthcare to fiber lasers and sensory technology. Traditionally composed of silica cores and claddings, optical fibers have seen an expansion in functionality and structure with the introduction of photonic crystals. These intrinsically micro-structured photonic crystal fibers, equipped with air holes of various diameters, offer unparalleled control over light propagation, enabling the design of task-specific optical fibers. Through the integration of insulators, semiconductors, and conductors, the scope of optical fibers has been broadened to include multi-material optical fibers. Plasmonic and metamaterial incorporations

---

Z. Meng · M. Chen (✉)
College of Biomedical Engineering & Instrument Science, Zhejiang University, Hangzhou, Zhejiang 310027, P. R. China
e-mail: mengxiaochen@zju.edu.cn

Z. Meng
e-mail: 12215017@zju.edu.cn

L. Wei (ed.), *Advanced Optical and Optoelectronic Fibers*, Advances in Optics and Optoelectronics, https://doi.org/10.1007/978-981-97-6218-7_6

also declare a new frontier for innovative possibilities in optical fibers applied across various electronic devices such as light-emitting diodes (LEDs), field-effect transistors (FETs), photodetectors, and solar cells. Unlike rigid planar optoelectronic devices, fiber-based alternatives offer unique qualities like flexibility, lightweight structure, and wearability, capturing spatial information otherwise unavailable. Based on these outstanding attributes, advanced optical and optoelectronic fiber devices have captured increasing attention, finding numerous applications in biological systems, thus signaling a pivotal turning point for future scientific and technological advances.

Achieving the optimal performance of optical and optoelectronic fibers is depended on their specific applications. In the case of epidermal sensors, the devices' capacity to conform to soft skin demands them to primarily be flexible and stretchable. However, when deployed as implantable devices, they encounter the substantial challenge of ensuring biocompatibility. Hydrogel-based optical and optoelectronic fibers, which are ascendant for their inherently biocompatible and stretchable properties, have thus emerged as a potential solution, gaining increasing attention. Nevertheless, the proven in vivo long-term stability of these fibers remains uncertain. So additional research for the comprehensive demonstration are necessary regarding the feasibility and safety of these fibers for implantable biomedical applications. Figure 1 demonstrates general strategies for designing advanced optical and optoelectronic fibers for bio-integrated systems.

This chapter provides a comprehensive overview of the materials commonly adopted in the fabrication of optical and optoelectronic fibers used in bio-integrated systems. In the subsequent section, we discuss three typical fabrication methods, namely thermal drawing, spinning, and 3D printing technology, which are widely

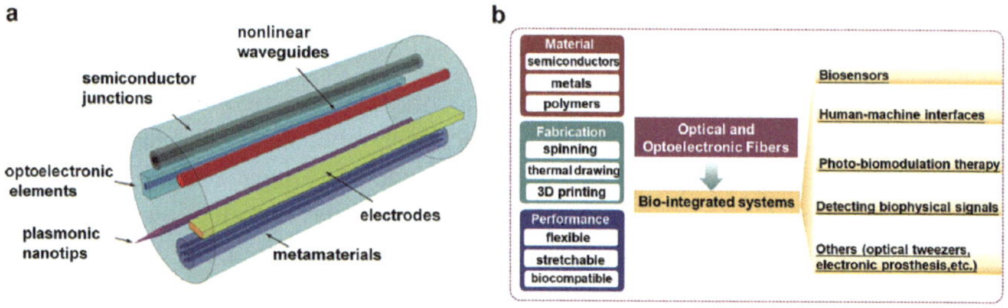

Fig. 1 The schematic of the fiber technology strategies for bio-integrated systems. **a** Designing of the multi-functional optoelectrical fibers, which are composed of multiple functional components, including nonlinear waveguides, semiconductor junctions, optoelectronic elements, metamaterials, etc. Adapted with permission [1]. Copyright 2015, Wiley. **b** The diagram to show the bio-integrated system elements based on optical and optoelectronic fibers. Material selections, fabrication methods, and expected performances are all important considerations when develop a fiber interface employed in bio-integrated system, such as detecting biophysical signals, biosensors, human–machine interfaces, photobiomodulation therapy, and others

employed in the manufacturing of these fibers. Finally, we highlight the specific applications of optical and optoelectronic fibers in bio-integrated systems, underscoring their potential in advancing biomedical research and healthcare.

2 Matrials, Fabrications and Applications

2.1 Materials and Fabrication Methods

2.1.1 Materials Selections in Bio-integrated System

Silica glass have been explored to manufacture optical and optoelectronic fibers for long, because of their excellently low attenuation and chemical inertness. As the requirements in wearable electronics and implantable devices arisen, researchers draw attentions from inorganic materials to develop organic polymers due to their higher flexibility, lower Young's modulus, higher strain limits, fracture toughness, lightweight, and impact resistance. Biocompatibility should be taken when optical and optoelectronic fibers are applied to biomedical research and healthcare. Material selection is a starting to offer the requested functionalities in bio-integrated system.

Materials Selections for Optical Fibers in Bio-integrated System

Traditional optical fibers rely on silica-based materials for manufacturing, but their rigidity and lack of extensibility can limit their use in contemporary electronics pursuing miniaturization and integration. Applications such as wearable technologies that adhere closely to soft skin necessitate flexibility and stretchability for long-term physiological monitoring and human–machine interfaces. Similarly, in implantable devices, ensuring high biocompatibility of optical and optoelectronic fibers becomes paramount to mitigate immune reaction with organs. Some researchers have employed hydrogel coatings to enhance the biocompatibility of these silica-based optical fibers, and materials that are biodegradable and bioresorbable are more preferred to negate the need for a secondary removal operation. Table 1 outlines a comprehensive summary of optical fiber materials and their corresponding applications in biosystems.

In the realm of biomedical applications, a broad spectrum of natural and synthetic polymers has been employed. Natural materials such as silk protein, cellulose, chitosan, and gelatin, favored for their inherent biocompatibility and processability, are commonly combined with other sensitive materials for monitoring physiological signals like temperature and breath. Hydrogels are another significant contributor to the field, drawing attention due to their high hydrophilicity and mechanical resemblance to biological tissues. They provide unique advantages for bio-integrated systems, particularly in guiding and manipulating light. This is mostly accredited to their low compatibility and refractive index mismatch with biological tissues when compared to solid-state materials [21]. Hydrogels made from poly(acrylamide)

Table 1 Commonly adopted optical fiber materials in bio-integrated system

Category	Materials	Application	References
Nature materials	Silk(core)	Waveguide	[2]
	Silk(cladding)	Drug release monitoring	[3]
	Cellulose(core)	Temperature sensing	[4]
	Chitosan(cladding)	Breathing monitoring	[5]
	Gelatin(cladding)	Breathing monitoring	[6]
	PEG (core)	Optogenetic	[7]
	PEG-poly(urethane) (PEG-PU) (cladding)	Optogenetics and electrophysiology	[8]
	PAAm	Optogenetic	[9]
	Poly(acrylamide-co-poly(ethylene glycol) diacrylate)(core)	Glucose sensing, Photomedicine	[10, 11]
	Poly(ethylene glycol diacrylate-co-acrylamide) (core)	Cancer therapy and brain optogenetic stimulation	[12]
Synthetic materials	PLA (core)	Biomarkers and cell apoptosis monitoring, Temperature sensing, photomedicine	[13, 14]
	PLLA & poly(L-actic-co-glycolic acid)(core)	Photobiomodulation	[15]
	Polycarbonate(core)	Photodynamic therapy	[16]
	PDMS (cladding)	Human–machine interfaces, temperature and movement monitoring	[17]
Synthetic materials (elastomer)	PDMS(coating)	Human–machine interfaces, blood pressure monitoring	[18, 19]
	PDMS(core, cladding)	Temperature sensing	[20]

(PAAm), poly(ethylene glycol) (PEG), and their copolymers have been fabricated int for strain monitoring, photo-biomodulation therapy and biomarkers monitoring. In addition to good biocompatibility and tunable biodegradability, polylactic acid (PLA) or poly(L-lactic acid) (PLLA) have high refractive indices ($\sim$1.46) and low extinction for optical waveguides, making them suitable for implantation in photobiomodulation therapy and in vivo physiological information monitoring. However, bioresorbable materials suffer from transmission and coupling losses due to their time-dependent biodegradability. Polydimethylsiloxane (PDMS), a commonly used elastomer, exhibits excellent optical properties [22], good elasticity, chemical inertness and thermosetting property, making it an ideal choice for human–machine interfaces.

Materials Selections for Optoelectronic Fibers in Bio-integrated System

Optoelectronic fibers have a broad range of materials used for their construction, comprising mainly of cladding, core, and an electrode, as summarized in Table 2 for bio-integrated systems. Initially, silica served as the primary cladding material, acting as a protective barrier for functional layers. However, in more recent advancements, borate glass and functional polymers have come into play. A diverse spectrum of semiconductor materials has been employed to manufacture the core layer of optoelectronic fibers, which can be categorized into five distinct types: (1) IV semiconductors and IV–IV alloys (Si, Ge, Si-Ge, and Sn). (2) III–V compound semiconductors (GaSb, InSb). (3) II–VI compound semiconductors (ZnSe). (4) Chalcogenide. (5) Polymers (polycarbonate (PC)). Compared to inorganic semiconductors, polymers-based optoelectronic devices offer advantages such as high flexibility, structure tunability and lightweight, making them suitable for wearable electronics. The conductive electrode layer facilitates signal transmission and power supply to the in-fiber optoelectronic devices. Metals such as gold, copper, tin and silver are commonly used as electrode due to their good conductivity and high melting temperatures. However, some metals have low viscosity and surface tension, which makes it difficult to maintain the fiber structure during the thermal drawing, leading to mixing with functional semiconductor materials and ultimately damaging the fiber structures and functions [23]. One potential solution is to introduce conductive polymers which have lower conductivity compared than metals but can be combined with conductive fillers (eg. carbon-based materials). Another approach is to deposit metallic nanowires on the surface of core fibers via dip-coating or high-pressure chemical vapor deposition.

Table 2 Commonly adopted functional core materials for optoelectronic fibers

Category	Group	Representative materials	References
Core (semiconductor)	IV semiconductors and IV-IV alloys	Si, Ge, Si-Ge, and Sn	[23]
	III-V compound semiconductors	GaSb, InSb	[24]
	II-VI compound semiconductors	ZnSe	[25]
	Chalcogenide	As-Se, As-S-Te	[26]
	Polymers	Polycarbonate	[27]
Electrode (conductor)	Metals	Ag nanowire, tin, gold, copper	[27]
	Polymers and carbon-based fillers	Polyethylene (CPE) and 5% graphite	[27]

2.1.2 Fabrication Techniques

Fabrication technologies determine the quality and architecture of optical and optoelectronic fibers. Typical fabrication methods include pressure-assisted filling, high pressure chemical vapor deposition (HPCVD), spinning, and thermal drawing. Some emerging techniques, like 3D printing, have the potential to achieve large-scale manufacturing and acquire different architectures. Recently, a variety of post-thermal drawing methods to functionalize fiber tips have been utilized, such as pattern transferring, photolithography, and self-assembly. It should be noticed that the adaptation of preparing technology and materials is particularly important. For example, similar thermal and chemical properties are required for thermal drawing technology.

Spinning

Spinning is a widely used method for producing polymer fibers, which are formed through extrusion. various spinning techniques have been developed, including wet spinning, dry spinning, electrospinning, microfluidic spinning, direct drawing and direct writing. In a recent study by Chen et al. [12], a light-triggered wet spinning process was employed to fabricate core-sheath hydrogel optical fibers (Fig. 2a). The core and sheath materials consisted of PEGDA and Ca-alginate, respectively. The spinning solution was prepared by dissolving a mixture of PEGDA and AAm monomer in water, and adding Na-alginate solution. The solution was then extruded through a spinning needle into a coagulate bath containing $CaCl_2$. During the spinning process, Na-alginate solution reacted with $CaCl_2$ through ionic crosslinking, while the core solution was crosslinked under UV illumination. Finally, the resulting fibers were collected on a bobbin. SEM images showed that the core-sheath fiber had a uniform structure with a diameter of several hundred micrometers.

Electrospinning is a method that leverages strong electrostatic force to produce smaller-diameter fibers with larger surface areas by extruding either a solution or melt. It is commonly used for creating nanofibers. Aspects including solution viscosity, polymer concentration, voltage levels, air humidity, working distance, and surface tension can greatly influence the fabrication process and the ultimate performance of the resulting fiber. Consequently, the production process is usually confined to a chamber with a stringent control over air temperature and humidity. Biocompatible polymers, including PLGA, PLLA, polyvinyl alcohol (PVA), cellulose acetate (CA), and silk, have been subject to electrospinning to generate optical and optoelectronic fibers suitable for biomedical applications. Kim and his colleagues [28] reported the preparation of stretchable photodetectors by electrospinning composite nanofibers. As depicted in Fig. 2b, Poly(vinylpyrrolidone) (PVP)/Methylammonium lead iodide ($MAPbI_3$) nanofibers were obtained on the surface of PDMS with Ag–Au alloy nanowires using an electrospinning technique and annealing process. The device was then encapsulated with thin PDMS film the device. Due to its high photocurrent, the device was utilized as a wearable image sensor.

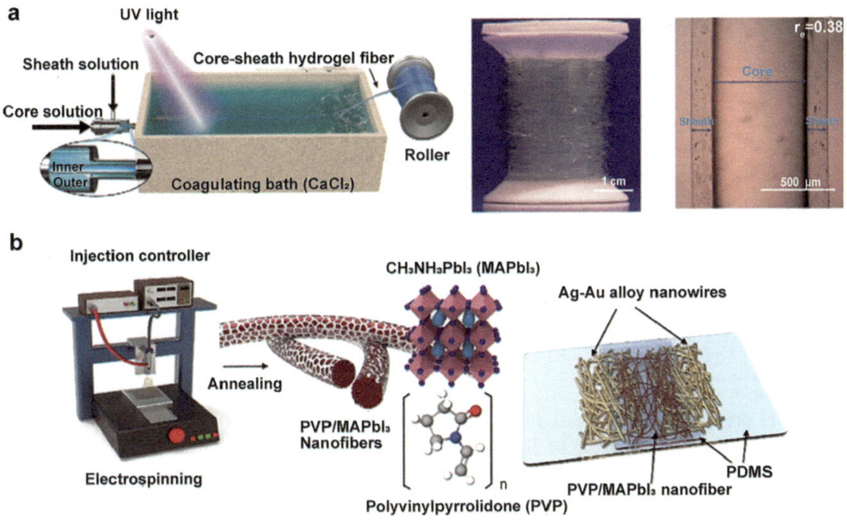

Fig. 2 Spinning technology to prepare the functional fibers. **a** Light-triggered wet spinning process. The core solution and sheath solution were injected into a coagulating bath, and fibers were formed through ionic crosslinking and polymerization under light irradiation. The resulting fibers were collected on a bobbin, and their morphology was characterized using SEM imaging, as shown in the figures. Reproduced under terms of the CC-BY license [12]. Copyright 2020, the Authors, published by Oxford University Press. **b** Schematic of electrospinning process used to fabricate PVP/CH3NH3PbI3 nanofibers, which were subsequently utilized as a photodetector in conjunction with Ag-Au alloy nanowires. Reprinted with permission [28]. Copyright 2022, American Chemical Society

3D Printing

Due to its rapid prototyping, cost-effectiveness, and ease of use, 3D printing is becoming an increasingly popular method for producing optical and optoelectronic fibers. Various 3D printing technologies, such as stereolithography (SLA), digital light projection (DLP), PolyJet, two-photon polymerization (2PP), fused deposition modeling (FDM) and direct ink writing (DIW), are shown in Fig. 3a. For biomedical applications, Zhuo and colleagues [29] utilized projection-suspended photocuring (PSP) 3D printing method to produce optical hydrogel fiber with high refractive-index (RI). Figure 3b illustrates that the PSP 3D printer utilizes digital light processing (DLP) for generating customized printing patterns and controlling the intensity distribution of curing light through computational coding. These patterns achieve submicron resolution via a microscope system, enabling the production of hollow fibers ranging from 1 to 800 μm in size. Compared to the SHF and SIHF, the RI of the GIHF was graded from the center of the core to the edge, and then connected to the cladding.

The combination of 3D printing and thermal drawing enables the production of functional fibers with arbitrary cross-section structures, as illustrated by the various preform designs shown in Fig. 3c. However, the anisotropic porosity inherent to the

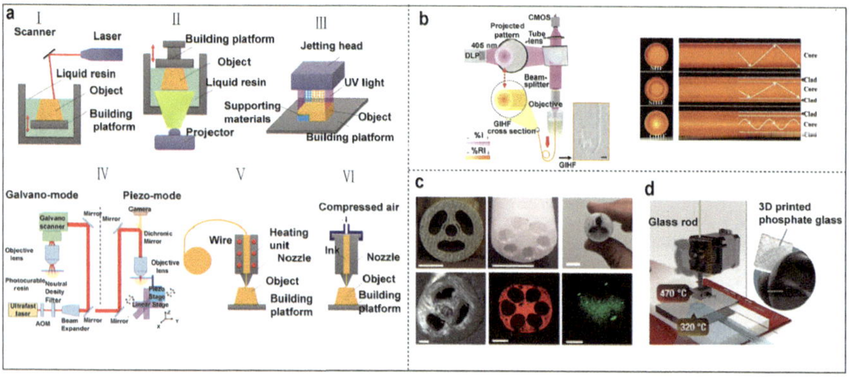

Fig. 3 3D printing technology as an emerging fabrication strategy for optical and optoelectrical fibers. **a** Schematic of four kinds of 3D printing technologies. I: SLA, II: DLP, III: PolyJet, IV: 2PP, V: FDM, VI: DIW. Reproduced under terms of Creative Common CC BY license [30]. Copyright 2018, the Authors, published by MDPI. Reproduced under terms of Creative Common CC BY license [31]. Copyright 2020, the Authors, published by MDPI. **b** Left: Schematic illustration of projection-suspended photocuring (PSP) 3D printing method to fabricate graded-index hydrogel fiber (GIHF). Right: phase-contrast images of single-core hydrogel fiber (SHF) (350 μm), step-index hydrogel fiber (SIHF) (250/350 μm), and GIHF (250/350 μm). Reproduced with permission [29]. Copyright 2023, Wiley. **c** 3D printed preforms with different architectures (scale bar = 1 cm). I: Preform with multiple irregular channels. II: Preform with six equitangetially spaced cylindrical channels. III: Preform with propeller-shaped channels. Reproduced under terms of the CC-BY license [32]. Copyright 2021, the Authors, published by Springer Nature. **d** Thermal drawn glass fiber as the feedback of 3D printing. Reproduced under terms of CC BY-NC-ND license [33]. Copyright 2020, the Authors, Published by Elsevier

3D printing process and the anisotropic thermomechanical and mechanical behavior of the preforms during the thermal drawing process have not been fully addressed. In contrast, a recent study utilized thermally drawn glass fiber as feedstock for FDM 3D printing of optically transparent structures, as shown in Fig. 3d. The results indicated that incorporating a dual material fiber into the 3D printing process can improve the mechanical properties of the materials [33].

Thermal Drawing

Thermal drawing technology is a traditional technology to fabricate optical fibers that enables the integration of materials with optical and photoelectric properties into fibers. Thus, it enables efficient preparation for flexible electronic devices with complex structures. The first step in thermal drawing involves preparing a macro preform, in which bulk materials are arranged in defined locations in well-defined geometric shapes (Fig. 4a). Many methods have been adopted to prepare preforms including HPCVD, rod-in-tube approach, sandwich consolidation and thin-film rolling. After heating the preform in the furnace for a period, the softened melt preform is extended to a length of up to tens of kilometers. By adjusting the drawing temperature, feeding speed of the preform, and drawing speed of the clamp, the diameter of the resulting fibers can be precisely controlled. Base on the above procedures,

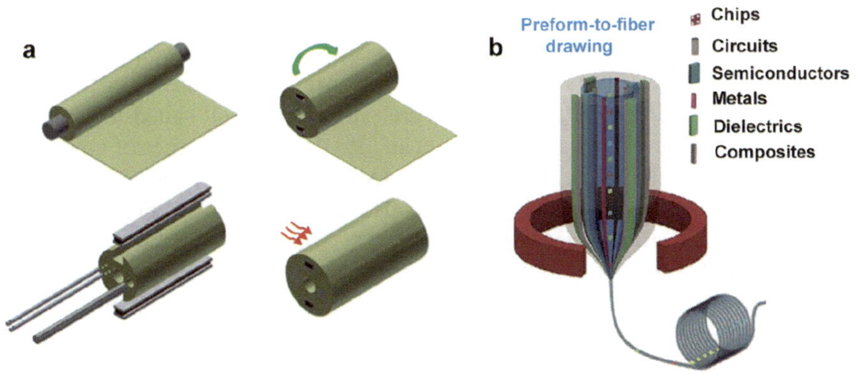

Fig. 4 Thermal drawing technology used to fabricate optical and optoelectronic fibers. **a** Schematic of fabricating the preform. Adapted with permission [24]. Copyright 2018, Wiley. **b** Schematic of thermal drawing process after preform fabrication and resulting optical or optoelectronic fibers containing functional material core and cladding. Reproduced under terms of the CC-BY license [34]. Copyright 2023, the Authors, published by Oxford University Press

opotoelectronic fiber devices can be fabricated through embedding insulators, semiconductors, and conductors into the fiber design (Fig. 4b). Compared to multi-stage fabrication methods that involve dip-coating or hydrothermal growth of each layer on the core fiber, thermal drawn optoelectronic fibers exhibit higher photoresponsivity (R_λ) and on/off ratio of photocurrent.

The thermal drawing technology offers numerous advantages such as superior uniformity, scalability, and high productivity. However, it is subject to limitations due to differences in thermal, mechanical and chemical properties of the various integrated materials. Therefore, randomly selected material combinations maybe not feasible to thermally drawn into fibers. For example, the cladding and core materials must have similar melting temperatures or glass temperatures. Moreover, at least one amorphous material is necessary for thermally drawn fiber, as amorphous materials can resist continuous deformation during the necking process from the preform to the resulting fibers. In addition, materials mismatch in thermal expansion properties and viscosity can result in fiber breakages and cracks.

2.2 *Practical Application of Advanced Optical and Optoelectronic Fibers for Bio-integrated Systems*

2.2.1 Human–Machine Interfaces

The field of human–machine interaction has become increasingly important in many areas, including human healthcare, robots, and virtual reality. Optical and optoelectronic fibers offer several advantages as human–machine interfaces, including immune to electromagnetic interference, high performance in humid environments, and the ability to sense multiple parameters simultaneously. Compared to other wearable electronic sensors that are based on resistance, capacitance or piezoelectricity, optical and optoelectronic fibers are particularly well-suited to large-area sensing applications. To enable this, Ma et al. [35] developed a series of novel interactively mechanochromic electronic textile (MET) sensors for visualized stretchable electronics. As shown in Fig. 5a, the electronic textile was obtained by integrating mechanochromic photonic elastomers (PE) with hierarchical-fiber-structured conductive polyester textiles (CPT). Notably, the MET sensor demonstrated a strong mechanochromic property and excellent reversibility. As shown in Fig. 5b, the orange-colored MET sensor's structural color gradually shifts from orange to green when stretched longitudinally from 0 to 20%. According to equation $\lambda = 2d(1-\nu\varepsilon_x)n_{eff}$, stretching the MET sensor horizontally reduced the lattice spacing vertically, leading to a decrease in wavelength λ. Due to the resistance change during the stretching process, the MET sensors demonstrated frequency-independent electromechanical properties (Fig. 5c). The MET sensor's excellent synchronous dual-signal response enabled its use as a visual interactive device. Color changes could be directly perceived by the human eye and quantified through spectroscopy, while electrical readings could be accurately recorded by electrical instruments, achieving direct human–machine interactive visualization (Fig. 5d). As a proof of concept, it was designed to monitor human joint movements by attaching it to the finger, wrist, elbow, knee, and neck, respectively (Fig. 5e, f). These findings demonstrate the potential of optical textiles electronics for advanced human–machine interactions.

In recent research on human–machine interaction, electronic readers have been a major focus; However, they may not always be accessible to the naked eyes. To address this issue, Yang and co-workers [36] reported the development of an optoelectronic intelligent fiber sensor (SOEFS) that is self-powered and provides feedback through both visible and electrical signals, based on the mechanoluminescent (ML)-triboelectric synergistic effect (Fig. 5g). The SOEFS is composed of a core of poly(vinylidene fluoride-trifluoroethylamine) (P(VDF-TrFE)) nanofibers and sheath of elastomer poly(styrene-b-(ethylene-co-butylene)-b-styrene (SEBS) and ML phosphors. This optoelectronic device exhibited significantly enhanced electrical output (100%) and optical output (30%) (Fig. 5h, i). When integrated into textiles, the optoelectronic bimodal SOEFS can create a synergy between vision and digital information during the stretching process (Fig. 5j). Furthermore, the SOEFS has been applied to various applications such as gesture recognition, AR interaction, robotic manipulation, and underwater information interaction (Fig. 5k–m).

2.2.2 Detecting Biophysical Signals

The integration of thermal-sensitive fluorescent materials into optical fibers has become a prevalent method for monitoring human body temperature. Guo and co-workers [20] fabricated stretchable optical fibers combined with up-conversion nanoparticles (UCNPs) NaYF4:Yb, Er. The optical fibers sensor demonstrated a

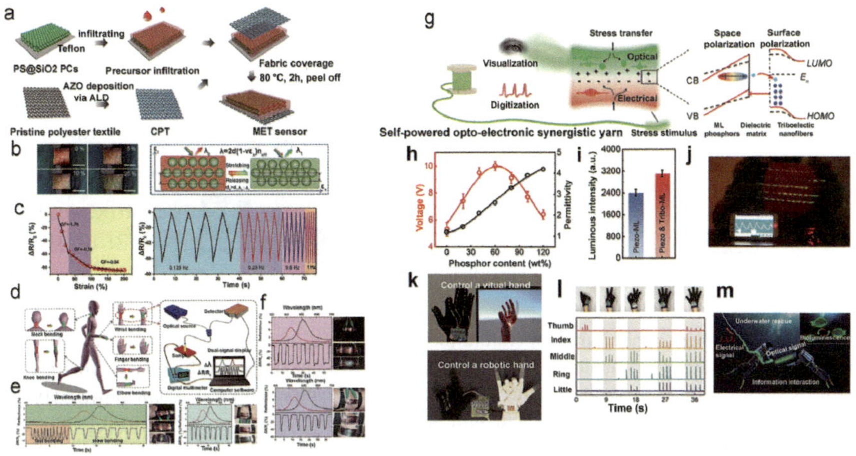

Fig. 5 Novel fiber devices used in human–machine interfaces. **a**–**f**: A mechanochromic electronic textile (MET) sensors for human–machine interface. **a** Schematic illustration of the preparation of the MET sensor. **b** Changes of structural color during stretching and releasing. **c** Electromechanical performance of the MET sensor. **d** Schematic illustration of the MET sensor as a visualized interactive device for detecting human joint movements. **e** Finger bending and wrist bending with different frequencies in real time. **f** Elbow bending and knee bending. Reproduced with permission [35]. Copyright 2021, Elsevier. **g**–**m**: Optoelectronic intelligent fiber sensor with visual–digital synergies. **g** The design of the SOEFS and optoelectronic synergy mechanism. **h** Voltage and permittivity in different content of phosphor. **i** Enhancement of luminescent intensity. **j** SOEFS enabled textiles with visual–digital synergies during stretching. SOEFS applied to AR interaction, robotic manipulation (**k**), gesture recognition (**l**) and underwater information interaction (**m**). Adapted with permission [36]. Copyright 2021, Wiley

temperature sensing range from 25 °C to 70 °C (Fig. 6a, b). When the sensor was attached to different positions of human body, it showed a fast response and the results were in accord with the normal temperature of a human body (interior, ≈37 °C; skin, 32–35 °C) (Fig. 6c). However, body movement induced strain has an adverse effect when stretchable optical fibers sensors are for wearable applications. Chen and co-workers [17] reported a stretchable optical fiber temperature sensor based on fluorescent nanoparticles and silicone elastomers. The excitation loss and fluorescence intensity of the sensor showed a linear response to changes in temperature and strain, which were utilized as sensing parameters to decouple the strain from the sensing signal.

Monitoring respiration is particularly crucial for individuals with respiratory conditions, heart failure, or those undergoing health rehabilitation training. Shrivastav and co-workers [5] reported an optical humidity sensor for human breath monitoring. The sensing material, chitosan, was deposited in the holes of a microstructured optical fiber, creating a Fabry-Pérot (FP) interferometer sensing platform (Fig. 6d). The working principle of the optical fiber sensor was based on volume changes of hygroscopic chitosan in different humidity, leading to the change of refractive index. Next, evanescent waves at the core-chitosan interface interacted in the FP

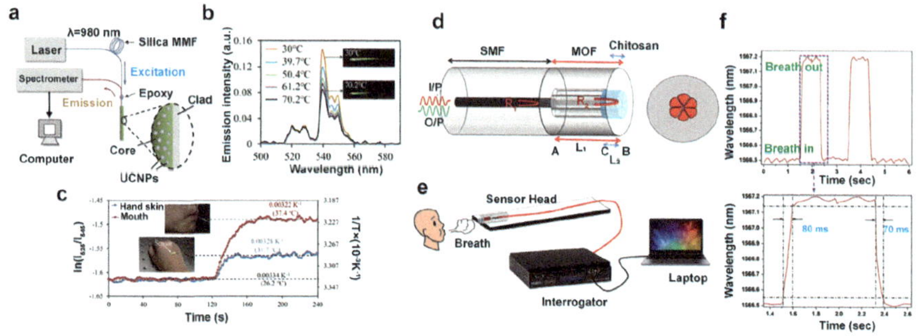

Fig. 6 Optical fibers applied in temperature monitoring (**a**–**c**) and respiratory (**d**–**f**) monitoring. **a** Schematic of the experimental setup for temperature sensing based on the optical fiber sensor with UCNPs. **b** Emission spectra of the sensor in different temperatures. **c** Temperature response of the sensor when attached to the hand skin or the mouth of a man. Adapted with permission [20]. Copyright 2019, Wiley. **d** Structural composition of the humidity sensor using a Fabry-Pérot interferometric configuration. **e** Experimental setup for respiratory monitoring. **f** Signal variation of the sensor when humans breathe in and breath out. Reproduced under terms of Creative Commons CC BY license [5]. Copyright 2020, the Authors, published by Springer Nature

cavity, resulting in the wavelength and power of reflected light to shift. Thus, the ambient humidity around the probe can be determined by monitoring the wavelength change of a specific tilt angle (Fig. 6e). From Fig. 6f, it's clearly seen that the wavelength difference of exhalation and inhalation could be distinguished. Additionally, the optical sensor showed ultrafast response and recovery time (80 ms and 70 ms, respectively).

Enhancing the comfort level of optical sensors necessitates a focus on streamlined patterning methods and the development of slimmer sensors. Zhu and co-workers [37] prepared an ultrathin optical sensor based on a wavy microfiber for blood pressure monitoring. The team employed a bottom-up approach to make the devices stretchable. The microfiber was tapered to several hundred nanometers to several micrometers and then was encapsulated into a pre-strained PDMS film (Fig. 7a). Upon release of the strain, the optical sensor and PDMS film formed a wave structure that demonstrated their flexibility, as shown by the sensor wrapped in a glass tube accompanied by red light (Fig. 7b). The optical sensor demonstrated mechanical response and adhesion to human skin, with a thickness comparable to an A4 sheet, ensuring (Fig. 7c). The blood pressure signal was defined by PAT, which represented the peak deviation value of the pulse waveform collected by the optical sensor and ECG waveform. The results obtained from healthy subjects showed that the blood pressure trend detected by the wave-like optical sensor was consistent with that measured using a cuff-based sphygmomanometer, highlighting the potential of the sensor for blood pressure monitoring (Fig. 7d).

Fiber optoelectronics technology is also being employed to detect physiological signals. The arrangement of micro-sized electrode patterns is crucial to make certain that the performance of fiber devices matches up to that of established planar devices. Kim and co-workers [38] introduced a fiber

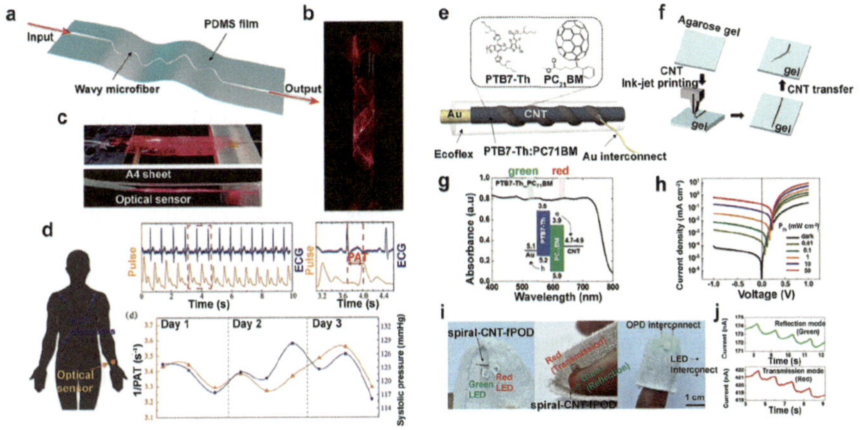

Fig. 7 Optical and optoelectronic fibers used to monitor in blood pressure (**a–d**) and heart rate (**e–j**). **a** Schematic of the wave-like optical sensor. **b** Photograph of flexibility verification of the as fabricated optical sensor. **c** Photograph of the ultrathin optical sensor compared with an A4 sheet. **d** Schematic of blood pressure monitoring and pulse signals analysis. **e** Schematic of device structure and materials composition of the fiber OPD. Adapted with permission [37]. Copyright 2021, Wiley. **f** Absorption spectrum of PTB7-Th:PC71BM in the range between 400 and 800 nm. **g** Photograph of the catheter wrapped on a glass rod. **h** Current density–voltage (J–V) curves of the fiber OPD. **i** Photographs of OPD with a red and a green LED as a PPG sensor. **j** PPG signals captured by the fiber OPD from reflection and transmission modes. Reprinted with permission [38]. Copyright 2020, American Chemical Society

organic photodiode (OPD) to monitor the heart rate. As shown in Fig. 7e, the fiber OPD composed of Au gate microfiber, Ecoflex elastomer as the protective layer, poly([2,6–4,8-di(5-ethylhexylthienyl)benzo[1,2-b;3,3-b]-dithiophene]{3-fluoro-2[(2-ethylhexyl)carbonyl]thieno[3,4-b]thiophenediyl)} (PTB7-Th) and [6]-phenyl-C71-butyric acid methyl ester (PC71BM) as the photoactive materials, and CNT microelectrode. The CNT microelectrode, prepared by inkjet-printing with a hydrogel template, was spirally wrapped on the fiber surface to ensure tight conformal contact (Fig. 7f). Figure 6g demonstrated that the photoactive layer had high absorbance in the green and red waves, which is the basis of combination with green and red LED as a sensor. The fiber OPD showed a high I_{light}/I_{dark} photocurrent ratio (Fig. 7h) and bending duration, which was important for wearable electronics. To create a photoplethysmography (PPG) sensor, the fiber OPD, as well as a green and a red LED, were integrated into a textile bandage (Fig. 7i). The fiber OPD received the red or green signals after transmission or reflection on the human tissue. Both mode measurements captured PPG signals with repeated systolic and diastolic peaks of the same period, as shown in Fig. 7j. The heart rate of the volunteer was estimably 63 bpm.

2.2.3 Biochemical Signals and Biosensors

The physiological pH is important in maintaining normal tissue function, which varies across different body areas. Notably, tumor sites always exhibit acidity, with changes in pH values specifying the wound healing status. Generally, pH is ascertained by placing organic dyes on a fiber tip. However, considering the health hazards associated with dyes, alternatives like biocompatible coatings and pH-sensitive carbon quantum dots have been utilized to mitigate toxicity to tissues or organs. Another challenge is dye-leaching and photobleaching lower the sensitivity and stability of the optical sensors [39]. Tang and co-workers [40] reported a dye-free optical fiber pH sensor. The sensor was U-shaped with a coated organic–inorganic composite membrane (Fig. 8a). The shape of the membrane changed since hydrogen bonds of the membrane changed when the sensor was in different pH solution and therefore the refractive index of the film varied. The sensor showed reversible response from 4.5 to 12.5 pH with accuracy of ± 0.2 pH compared with a commercial probe (Fig. 8b, c). For biomedical application, the sensor was employed to monitor the pH of human serum (Fig. 8d), which demonstrated the proposed sensor could record the dynamic changes while adjusting the pH value of serum from about 7.7 to 6.4.

Various glucose monitoring systems, have been utilized for short-term glucose detection, including implantable electrochemical sensors, but they come with certain limitations, such as signal drift due to enzymatic reactions and accuracy constraints [42]. Hydrogel-based optical fibers show good potential because of their biocompatibility, flexibility and capability to be combined with functional groups for glucose sensing. In a study by Mohamed Elsherif and co-workers [10], hydrogel optical fibers were prepared using poly(acrylamide-co-poly(ethylene glycol) diacrylate) functionalized with phenylboronic acid as the core layer, and Ca alginate as the cladding layer (Fig. 8e). The complexation of phenylboronic acid and cis-diol groups of glucose allowed the diameter of the hydrogel fiber to change reversibly (Fig. 8f). Gelatin matrices were utilized as phantom tissues in vitro to visualize propagation of light throughout the hydrogel optical fibers (Fig. 8g). Results showed the light attenuation was less than 10% when the optical fibers were inserted into gelatin matrices under a laser lamination. Moreover, the fabricated hydrogel fiber could deliver light into the targeted area when it was injected into porcine tissue ex vivo (Fig. 8h). However, despite these promising results, few optical sensors have been implanted in vivo for long-term glucose monitoring. Designing effective implantable optical sensors for glucose monitoring requires careful consideration of the device configuration, especially the configuration of electrode micropatterns, to ensure that the performance of fiber devices is comparable to established planar devices.

The global pandemic of Corona Virus Disease 2019 (COVID-19) has brought significant challenges to the field of medical diagnosis. Although reverse transcription polymerase chain reaction (RT-PCR) is widely used to detect SARS-CoV-2 infection, it has limitations including high false-negative rates and complexity. As an alternative solution, biosensors and spectroscopy based on optical fibers have been proposed. Nunzio Cennamo and co-workers [41] developed a plasmonic D-shaped

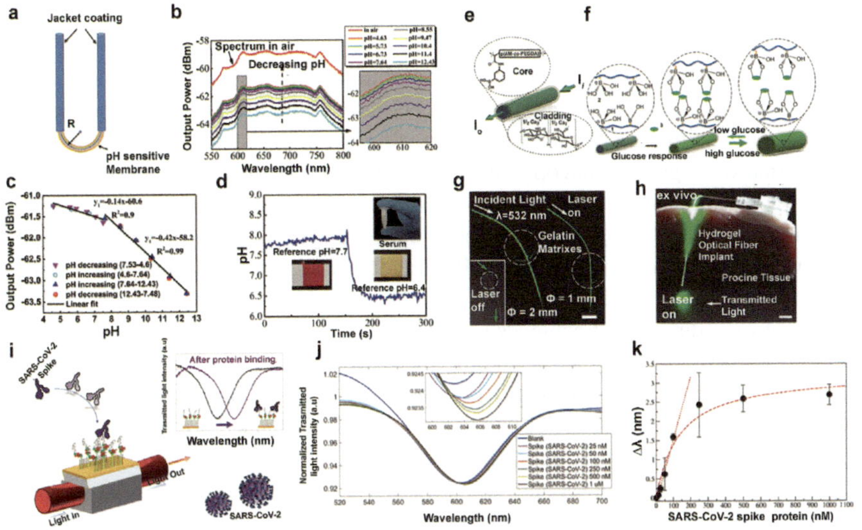

Fig. 8 Optical and optoelectronic fibers applied in biosensors. **a**–**d**: Optical fibers in pH sensing. **a** Schematic of the U-shaped optical fiber sensor with a pH sensitive membrane. **b** The spectral response of the optical fiber sensor in different pH. **c** Linear fitting curves. **d** pH measurement of human serum with the sensor. Reproduced under terms of Creative Commons Attribution 4.0 License [40]. Copyright 2021, the Authors, published by IEEE Xplore. **e**–**h**: Hydrogel-based optical fibers in glucose monitoring. **e** Schematic of structural composition of the hydrogel optical fibers. **f** Mechanism of glucose sensitive of the sensor. **g** Light transmission of hydrogel optical fibers through gelatin matrixes. Scale bar = 1 cm. The inset showed the hydrogel optical fiber was sandwiched in gelatin when the laser was off. **h** Light-guiding in porcine tissue. Scale bar = 3 mm. Reproduced under the terms of the Creative Commons CC BY license [10]. Copyright 2017, the Authors, published by Wiley. **i**–**k**: SARS-CoV-2 spike protein detection. **i** Schematic of interaction of SARS-CoV-2 protein and aptameric sequence. **j** Transmission spectra of protein concentration (25–1000 nM). **k** Resonance shift ($\Delta\lambda$) in different spike protein concentration. Reproduced with permission [41]. Copyright 2021, Elsevier

plastic optical fiber that can detect the spike protein of SARS-CoV-2 virus with high sensitivity and a protein concentration response range of 25 to 1000 nM, and a detection limit of 36.7 nM (Fig. 8j, k). They achieve this by immobilizing an aptamer on a gold surface via polyethylene glycol, which changes the light intensity upon interaction with the virus protein and aptameric sequence (Fig. 8i). Meanwhile, Wu and his team [43] developed a portable smartphone-based optical fiber that integrates an immunosensing system for diagnosing COVID-19. The device is approximately 108 mm × 158 mm × 150 mm in size and can analyze serum samples from clinically diagnosed COVID-19 patients and healthy volunteers. The device showed a sensitivity of 72% and a specificity of 100%, which was highly correlated with those obtained using commercial ELISA kits and simple Western analysis. These optical fiber based biosensors offer a promising platform for fast, accurate, and cost-effective COVID-19 detection, which could significantly improve the current state of diagnostics.

2.2.4 Photobiomodulation (PBM) Therapy

Emerging as a promising treatment, Photobiomodulation (PBM) therapy offers potential benefits for tissue regeneration, photodynamic therapy, and optogenetics, primarily by facilitating light delivery to specific areas in vivo. Optogenetics employs a combination of genetic and optical tactics to manipulate the excitatory or inhibitory states of neurons and record electrophysiological signals from targeted regions. Through the use of transgene technology, photosensitive proteins can be specifically instantiated on neuron membranes, allowing them to react to various wavelengths and influence neuronal states. Park and colleagues [27] have developed a flexible optoelectronic device that achieves one-step optogenetics. This thermal drawn fiber device comprises an optical waveguide made of polycarbonate (PC) (core)/cyclic olefin copolymer (COC) (cladding), six electrodes made of conductive polyethylene (CPE) and 5% graphite (gCPE), and two microfluidic channels (Fig. 9a). The opsin genes were delivered into specific sites of a mouse's brain through the microfluidic channels, thus avoiding a second surgery. After 2 weeks of implantation, the optoelectronic device enabled the expression of the CHR2 photosensitive protein (protein to excite neurons) (Fig. 9b). Subsequently, the device captured electrophysiological signals of optically evoked potentials via light transmission of the PC/COC waveguide (Fig. 9c). The device's stable optical stimulation capability was demonstrated up to 3 months by recording the potentials (Fig. 9d). Due to its all-polymer composition, the optoelectronic fiber device has superior biocompatibility compared to steel microwire (Fig. 9e).

The rigid light delivery devices made of silica can damage soft tissues and impede natural animal behaviors [45]. Liu and co-workers [44] developed soft and stretchable fatigue-resistant hydrogel optical fibers that allow for optogenetic modulation of peripheral nerves in naturally behaving animals during continuous movement. The hydrogel-based device includes a short silica segment to prevent hydrogel dehydration in air, a polyvinyl alcohol (PVA) hydrogel fiber with a core-cladding structure, and a hydrogel cuff attached to the targeted nerve (Fig. 9f). The cladding hydrogel with PVA nanocrystalline domain enhanced hydrogel core's refractive index and minimized light loss. The hydrogel fibers showed low light attenuation coefficients of 1.07 dB cm^{-1} at 473 nm and 0.53 dB cm^{-1} at 589 nm in the fully hydrated state (Fig. 9g). After more than 1,000 stretch cycles, the hydrogel fibers remain light transmission above 87% (Fig. 9h). Unlike silicone fibers that restricted arterial displacement, hydrogel fibers followed tissue deformation (Fig. 9i). When optical stimuli was applied to tibialis anterior or gastrocnemius muscles (Fig. 9j), the EMG signal were stronger (Fig. 9k), indicating good potential of the hydrogel in peripheral nerve optogenetics during locomotion.

By inserting optical fibers into the cavity of human organs, photodynamic therapy has been used clinically for the non-invasive treatment of local tumors. To improve therapy efficiency and reduce the side effects of photosensitizers, researchers have opted for intratumoral injection instead of intravenous injection. However, instillation into targeted cancer sites does not prevent photosensitizers from spreading to distant organs. To achieve local controlled release, researchers have explored

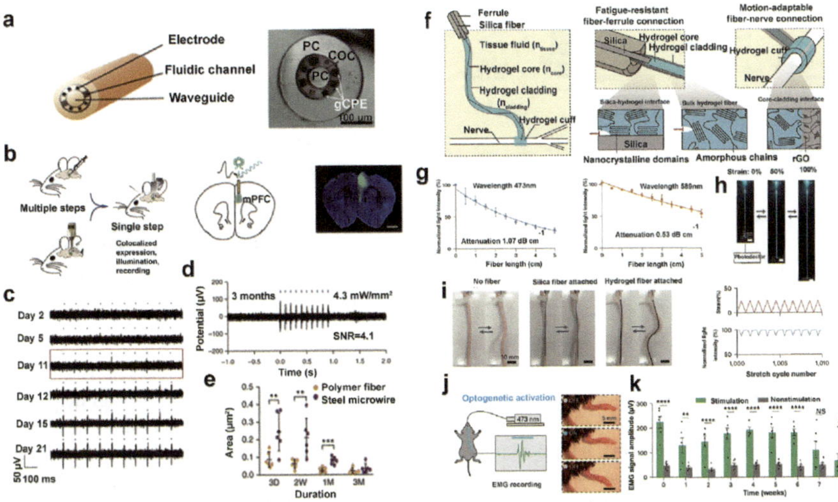

Fig. 9 Optical and optoelectronic fibers in PBM therapy. **a**–**e**: An implantable optoelectronic fiber device for optogenetics. **a** A schematic and cross-sectional microscope image of the multifunctional optoelectronic fiber. **b** Left: a schematic comparison between traditional two-step surgery and one-step surgery in this research for optogenetics. Middle: a working process illustration of the fiber probe including viral delivery, optical stimulation and electrical recording. Right: Fluorescent image of expression of ChR2-eYFP for two weeks after implantation. **c** Optically evoked potentials recording from day 2 to day 21 by the fiber. **d** Optically evoked potentials recording after 3 months. **e** Average fluorescence area of glial scarring of the polymer fiber and steel microwire. Reproduced with permission [27]. Copyright 2017, Springer Nature. **f**–**k**: Hydrogel optical fibers for peripheral nerve optogenetics during locomotion. **f** Schematic of the structure of a hydrogel fiber. **g** Plot of normalized light intensity versus fiber length at 473 nm and 589 nm. **h** The light intensity of the hydrogel fiber under the strain. **i** Images of a porcine artery with no fiber, an artery with a silica fiber attached and an artery with a hydrogel fiber attached. **j** Schematic showing the optogenetic stimulation controlled by optical pulses from a 473-nm light source and recorded by EMG. Scale bars, 5 mm. **k** Plot of EMG signal amplitude from week 0 before implantation to week 8 after implantation with and without optogenetic stimulation. Adapted with permission [44]. Copyright 2023, Spring Nature

coating photosensitizers on the surface of implantable optical fibers. This approach has proven effective in prolonging the retention of immunotherapeutics, thereby enhancing anti-tumor efficacy when combined with immunotherapy [16].

2.2.5 Others

Beyond those mentioned above, advanced fiber technology has also facilitated the advent of other groundbreaking tools. Among them, optical tweezers stand out as an invaluable instrument for optical trapping and manipulation. It has been applied broadly in cell trapping, labeling, analysis, and assembly to investigate cell–cell interactions. The capacity to arrange randomly distributed cells into systematic structures

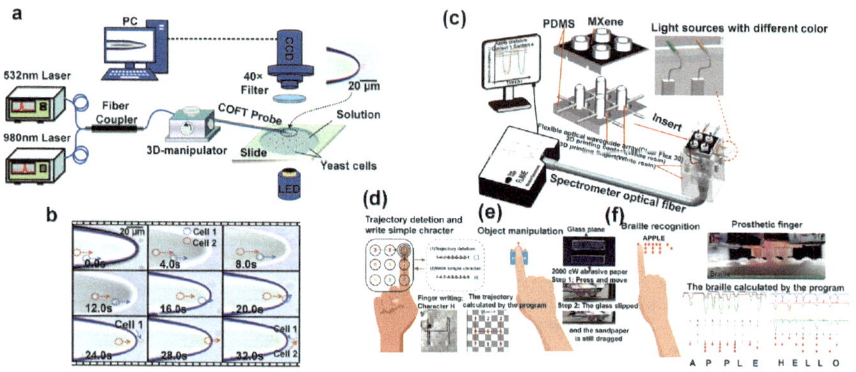

Fig. 10 Optical and optoelectronic fibers-based devices toward optical tweezers and electronic prosthetics. **a** Schematic of COFT system for optical manipulation. The inset shows a trapped yeast cell by the COFT probe. **b** Optical microscopy images of the micromanipulation of two cells. Reproduced with permission [46]. Copyright 2022, Royal Society of Chemistry. **c** Schematic of structure of the prosthetic finger. Application demonstration of **d** Writing character H. **e** Manipulating objects and comparing roughness. **f** Braille recognition. Reproduced with permission [47]. Copyright 2023, Royal Society of Chemistry

and arrays using optical forces offers further potential for biomedical fields such as targeted therapy, drug delivery, and microfluidic chip engineering. Deng at el. [46] recently reported the development of a capillary optical fiber tweezer (COFT) for the ballistic transport and trapping of yeast cells (Fig. 10a). The COFT is capable of capturing over two cells simultaneously during the trapping process (Fig. 10b). However, biocompatibility is a crucial consideration for optical tweezers, as they can cause mechanical damage when they are implanted into an organism.

Zhao et al. [47] reported a 3D printing prosthetic finger with flexible chromatic optical waveguides implanted at the fingertip. The core component of this multi-functional system is an elastic optical waveguide array with a grid structure. The optical waveguide array consists of four optical fibers. Additionally, MXene coatings attached on the PDMS made the system temperature sensitive (Fig. 10c). The artificial finger realized the integrated detection of multiple sensing parameters, including object contour, hardness, slip direction, velocity, and temperature. The final demonstration used this finger to writing a character, manipulating objects, feel roughness and recognize Braille (Fig. 10d–f).

3 Conclusion

Advanced fiber technologies have exhibited substantial potential for application in wearable sensors, human–machine interfaces, and implantable devices, facilitated by appropriate materials and fabrication methods. However, fiber devices still confront several challenges when integrated into biological systems. Firstly, specific to each application, the biomaterials incorporated in optical and optoelectronic fibers must be selected not just for biocompatibility, but also for the requisite mechanical properties and degradation time requirement. Secondly, fiber instrumentation is prone to higher optical attenuation due to bending, thus restrict their effective range for short-distance communication. Thirdly, the effective miniaturization of implantable devices integrated with additional electronic components remains a significant challenge. Consequently, future research about optical and optoelectronic fibers should dedicate focus to the exploration of materials and fabrication methods that cater to optical transmission and adaptability with tissues. The utilization of AI techniques can possibly address these challenges, assisting in tasks such as materials library selection and outcome prediction, analytical data processing, and the conversion of optical signals into meaningful parameters.

References

1. Alexander Schmidt M, Argyros A, Sorin F. Hybrid optical fibers—an innovative platform for in-fiber photonic devices. Adv Opt Mater. 2015;4(1):13–36.
2. Qiao X, Qian Z, Li J, Sun H, Han Y, Xia X, Zhou J, Wang C, Wang Y, Wang C. Synthetic engineering of spider silk fiber as implantable optical waveguides for low-loss light guiding. ACS Appl Mater Interfaces. 2017;9(17):14665–76.
3. Liu M, Zhang Y, Liu K, Zhang G, Mao Y, Chen L, Peng Y, Tao TH. Biomimicking antibacterial opto-electro sensing sutures made of regenerated silk proteins. Adv Mater. 2021;33(1): e2004733.
4. Skwierczynska M, Stopikowska N, Kulpinski P, Klonowska M, Lis S, Runowski M. Ratiometric upconversion temperature sensor based on cellulose fibers modified with yttrium fluoride nanoparticles. Nanomaterials (Basel). 2022;12(11).
5. Shrivastav AM, Gunawardena DS, Liu Z, Tam HY. Microstructured optical fiber based Fabry-Perot interferometer as a humidity sensor utilizing chitosan polymeric matrix for breath monitoring. Sci Rep. 2020;10(1):6002.
6. Yi Y, Jiang Y, Zhao H, Brambilla G, Fan Y, Wang P. High-performance ultrafast humidity sensor based on microknot resonator-assisted mach-zehnder for monitoring human breath. ACS Sens. 2020;5(11):3404–10.
7. Feng J, Zheng Y, Bhusari S, Villiou M, Pearson S, Campo A. Printed degradable optical waveguides for guiding light into tissue. Adv Funct Mater. 2020;30(45).
8. Tabet A, Antonini MJ, Sahasrabudhe A, Park J, Rosenfeld D, Koehler F, Yuk H, Hanson S, Stinson J, Stok M, Zhao X, Wang C, Anikeeva P. Modular integration of hydrogel neural interfaces. ACS Cent Sci. 2021;7(9):1516–23.
9. Wang L, Zhong C, Ke D, Ye F, Tu J, Wang L, Lu Y. Ultrasoft and highly stretchable hydrogel optical fibers for in vivo optogenetic modulations. Adv Opt Mater. 2018;6(16).
10. Yetisen AK, Jiang N, Fallahi A, Montelongo Y, Ruiz-Esparza GU, Tamayol A, Zhang YS, Mahmood I, Yang SA, Kim KS, Butt H, Khademhosseini A, Yun SH. Glucose-sensitive hydrogel optical fibers functionalized with phenylboronic acid. Adv Mater. 2017;29(15).
11. Jiang N, Ahmed R, Rifat AA, Guo J, Yin Y, Montelongo Y, Butt H, Yetisen AK. Functionalized flexible soft polymer optical fibers for laser photomedicine. Adv Opt Mater. 2018;6 (3).
12. Chen G, Wang G, Tan X, Hou K, Meng Q, Zhao P, Wang S, Zhang J, Zhou Z, Chen T, Cheng Y, Hsiao BS, Reichmanis E, Zhu M. Integrated dynamic wet spinning of core-sheath hydrogel fibers for optical-to-brain/tissue communications. Natl Sci Rev. 2021;8(9), nwaa209.
13. Zhang W, Liu P, Yang G, Lei H. Single polylactic acid nanowire for highly sensitive and multifunctional optical biosensing. ACS Appl Mater Interfaces. 2021;13(24):27983–90.

14. Zhang W, Huang X, Liu W, Gao Z, Zhong L, Qin Y, Li B, Li J. Semiconductor plasmon enhanced upconversion toward a flexible temperature sensor. ACS Appl Mater Interfaces. 2023;15(3):4469–76.
15. Jiang Y, Qi W, Zhang Q, Liu H, Zhang J, Du N, Nazempour R, Su Y, Fu R, Zhang K, Lyu P, Dong F, Yin L, Sheng X, Wang Y. Green light-based photobiomodulation with an implantable and biodegradable fiber for bone regeneration. Small Methods. 2020;4(7).
16. Chin AL, Jiang S, Jang E, Niu L, Li L, Jia X, Tong R. Implantable optical fibers for immunotherapeutics delivery and tumor impedance measurement. Nat Commun. 2021;12(1):5138.
17. Chen M, He Y, Liang H, Zhou H, Wang X, Heng X, Zhang Z, Gan J, Yang Z. Stretchable and strain-decoupled fluorescent optical fiber sensor for body temperature and movement monitoring. ACS Photonics. 2022;9(4):1415–24.
18. Li T, Su Y, Chen F, Liao X, Wu Q, Kang Y, Tan Y, Zhou Z. A skin-like and highly stretchable optical fiber sensor with the hybrid coding of wavelength–light intensity. Adv Intell Syst. 2021;4(4).
19. Pang Y-N, Liu B, Liu J, Wan S-P, Wu T, Yuan J, Xin X, He X-D, Wu Q. Singlemode-multimode-singlemode optical fiber sensor for accurate blood pressure monitoring. J Lightwave Technol. 2022;40(13):4443–50.
20. Guo J, Zhou B, Yang C, Dai Q, Kong L. Stretchable and temperature-sensitive polymer optical fibers for wearable health monitoring. Adv Funct Mater. 2019;29(33).
21. Guimaraes CF, Ahmed R, Marques AP, Reis RL, Demirci U. Engineering hydrogel-based biomedical photonics: design, fabrication, and applications. Adv Mater. 2021;33(23): e2006582.
22. Prajzler V, Neruda M, Nekvindová P. Flexible multimode polydimethyl-diphenylsiloxane optical planar waveguides. J Mater Sci: Mater Electron. 2018;29(7):5878–84.
23. Zhang J, Wang Z, Wang Z, Wei L. Advanced multi-material optoelectronic fibers: a review. J Lightwave Technol. 2021;39(12):3836–45.
24. Yan W, Page A, Nguyen-Dang T, Qu Y, Sordo F, Wei L, Sorin F. Advanced multimaterial electronic and optoelectronic fibers and textiles. Adv Mater. 2019;31(1): e1802348.
25. Seyedin S, Carey T, Arbab A, Eskandarian L, Bohm S, Kim JM, Torrisi F. Fibre electronics: towards scaled-up manufacturing of integrated e-textile systems. Nanoscale. 2021;13(30):12818–47.
26. Xu H, Yin L, Liu C, Sheng X, Zhao N. Recent advances in biointegrated optoelectronic devices. Adv Mater. 2018;e1800156.
27. Park S, Guo Y, Jia X, Choe HK, Grena B, Kang J, Park J, Lu C, Canales A, Chen R, Yim YS, Choi GB, Fink Y, Anikeeva P. One-step optogenetics with multifunctional flexible polymer fibers. Nat Neurosci. 2017;20(4):612–9.
28. Kim H-J, Oh H, Kim T, Kim D, Park M. Stretchable photodetectors based on electrospun polymer/perovskite composite nanofibers. ACS Applied Nano Mater. 2022;5(1):1308–16.
29. Zhuo X, Han Y, Bian Y, Xu A, Shen H. A promising optical bio-interface: graded-index hydrogel fiber. Adv Opt Mater. 2023;12(4).
30. Liu C, Huang N, Xu F, Tong J, Chen Z, Gui X, Fu Y, Lao C. 3D printing technologies for flexible tactile sensors toward wearable electronics and electronic skin. Polymers (Basel). 2018;10 (6).
31. Koo S. Advanced micro-actuator/robot fabrication using ultrafast laser direct writing and its remote control. Appl Sci. 2020;10(23).
32. van der Elst L, Faccini de Lima C, Gokce Kurtoglu M, Koraganji VN, Zheng M, Gumennik A. 3D printing in fiber-device technology. Adv Fiber Mater. 2021;3(2), 59–75.
33. Zaki RM, Strutynski C, Kaser S, Bernard D, Hauss G, Faessel M, Sabatier J, Canioni L, Messaddeq Y, Danto S, Cardinal T. Direct 3D-printing of phosphate glass by fused deposition modeling. Mater Design. 2020;194.
34. Qian S, Liu M, Dou Y, Fink Y, Yan W. A 'Moore's law' for fibers enables intelligent fabrics. Natl Sci Rev. 2023;10(1), nwac202.
35. Zhao K, Cao X, Alsaid Y, Cheng J, Wang Y, Zhao Y, He X, Zhang S, Niu W. Interactively mechanochromic electronic textile sensor with rapid and durable electrical/optical response for visualized stretchable electronics. Chem Eng J. 2021;426.

36. Yang W, Gong W, Gu W, Liu Z, Hou C, Li Y, Zhang Q, Wang H. Self-Powered Interactive Fiber Electronics with Visual-Digital Synergies. Adv Mater. 2021;33(45): e2104681.
37. Zhu H. T, Zhan L. W, Dai Q, Xu B, Chen Y, Lu YQ, Xu F. Self-assembled wavy optical microfiber for stretchable wearable sensor. Adv Opt Mater. 2021;9(11).
38. Kim H, Kang TH, Ahn J, Han H, Park S, Kim SJ, Park MC, Paik SH, Hwang DK, Yi H, Lim JA. Spirally wrapped carbon nanotube microelectrodes for fiber optoelectronic devices beyond geometrical limitations toward smart wearable e-textile applications. ACS Nano. 2020;14(12):17213–23.
39. Rezapour Sarabi M, Jiang N, Ozturk E, Yetisen AK, Tasoglu S. Biomedical optical fibers. Lab Chip. 2021;21(4):627–40.
40. Tang Z, Gomez D, He C, Korposh S, Morgan SP, Correia R, Hayes-Gill B, Setchfield K, Liu L. A U-shape fibre-optic ph sensor based on hydrogen bonding of ethyl cellulose with a Sol-Gel matrix. J Lightwave Technol. 2021;39(5):1557–64.
41. Cennamo N, Pasquardini L, Arcadio F, Lunelli L, Vanzetti L, Carafa V, Altucci L, Zeni L. SARS-CoV-2 spike protein detection through a plasmonic D-shaped plastic optical fiber aptasensor. Talanta. 2021;233: 122532.
42. Gupta BD, Pathak A, Shrivastav AM. Optical biomedical diagnostics using lab-on-fiber technology: a review. Photonics. 2022;9(2).
43. Wu Z, Wang C, Liu B, Liang C, Lu J, Li J, Tang X, Li C, Li T. Smartphone-based high-throughput fiber-integrated immunosensing system for point-of-care testing of the SARS-CoV-2 nucleocapsid protein. ACS Sens. 2022;7(7):1985–95.
44. Liu X, Rao S, Chen W, Felix K, Ni J, Sahasrabudhe A, Lin S, Wang Q, Liu Y, He Z, Xu J, Huang S, Hong E, Yau T, Anikeeva P, Zhao X. Fatigue-resistant hydrogel optical fibers enable peripheral nerve optogenetics during locomotion. Nat Methods. 2023;20(11):1802–9.
45. Liu X, Liu J, Lin S, Zhao X. Hydrogel machines. Mater Today. 2020;36:102–24.
46. Deng H, Chen D, Wang R, Li F, Luo Z, Deng S, Yin J, Yu L, Zhang W, Yuan L. Fiber-integrated optical tweezers for ballistic transport and trapping yeast cells. Nanoscale. 2022;14(18):6941–8.
47. Zhou J, Fu C, Fang J, Shang K, Pu X, Zhang Y, Jiang Z, Lu X, He C, Jia L, Yao Y, Qian L, Yang T. Prosthetic finger for fingertip tactile sensing via flexible chromatic optical waveguides. Mater Horiz. 2023;10(11):4940–51.

Chapter 7
Fiber X-Ray Detectors

Xi Chen, Shengpeng Hu, Mingzhu Hu, Yumeng Wang, Chunlei Yang, and Ming Chen

Abstract X-ray has become an essential in human's daily life, extensively used in many fields, like scientific research, medical imaging and nondestructive testing. The X-ray detector, as one of the most momentous modules of a related equipment, plays a vitally important role in the application of X-ray. At present, X-ray detectors are mainly used for two purposes, dosage detection (single pixel detectors) and imaging (detector arrays). Compared with other counterparts, like flat panel detectors, the fiber ones possess unique superiority due to the geometric characteristic. In dosage detection, the fiber structure results in miniaturization, flexibility and remote measurement. As to imaging, the high length-to-diameter ratio of single fiber leads to not only preeminent performance of suppressed cross-scattering of light, enhanced radiative absorption and improved spatial resolution, also the facile preparation.

Xi Chen, Shengpeng Hu—*These authors contributed equally.

X. Chen · S. Hu · M. Hu · Y. Wang · C. Yang · M. Chen (✉)
Center for Photonics Information and Energy Materials, Shenzhen Institute of Advanced Technology, Chinese Academy of Sciences, Shenzhen 518055, People's Republic of China
e-mail: ming.chen2@siat.ac.cn

X. Chen
e-mail: xichen@siat.ac.cn

S. Hu
e-mail: sp.hu@siat.ac.cn

M. Hu
e-mail: mz.hu@siat.ac.cn

Y. Wang
e-mail: ym.wang5@siat.ac.cn

C. Yang
e-mail: cl.yang@siat.ac.cn

X. Chen · C. Yang · M. Chen
University of Chinese Academy of Sciences, Beijing 100049, People's Republic of China

S. Hu · M. Hu · Y. Wang
Nano Science and Technology Institute, University of Science and Technology of China, Suzhou 215123, People's Republic of China

L. Wei (ed.), *Advanced Optical and Optoelectronic Fibers*, Advances in Optics and Optoelectronics, https://doi.org/10.1007/978-981-97-6218-7_7

According to the different interactions between fiber materials and X-ray photons, detection mechanisms of fiber detectors could be classified into indirect and direct type. And the preparation methods of fiber detectors are relatively simple with the participation of fiber optical or template. Besides, the performance parameters of fiber detectors in previous researches are summarized and analyzed. And the remain challenges for fiber detectors and potential solutions are also discussed.

Keywords Fibers · X-ray detectors · X-ray imaging · High spatial resolution · Low detection limit

1 Introduction

The discovery of X-ray in 1895 was one of the most significant milestones in human history, which has opened up a new and vast horizon of scope in medicine, physics, industry and many other dimensions [1, 2]. For instance, X-ray imaging examination, like projection radiography and fluoroscopy and computed tomography (CT), plays a quite important role in clinical diagnosis. What's more, X-ray could also be applicated in radio therapy and pharmaceutical analysis [3]. In recent years, X-ray diffraction (XRD), small-angle X-ray scattering (SAXS), X-ray photoelectron spectroscopy (XPS), energy dispersive X-ray spectrometry (EDS), X-ray absorption spectroscopy (XAS) and many other characterization techniques based on X-ray have been acting as crucial analysis tools in scientific research [4]. At the same time, X-ray testing is an essential technique of industry testing method, which could detect inner invisible defects of the tested subject exactly but without any physical damage to it [5]. In brief, X-ray has been extensively applicated and continuously making sense in every facet of human production and living.

Nevertheless, with the improvement of science and technology, the risk of X-ray to human health caused by ionization has gradually caught public attention as well [6]. Besides lead shield, radiation dosage control is another effective gateway to reduce the harm of electromagnetic radiation [7]. However, under the present condition, higher radiation exposure has to be applicated for better imaging quality [8]. As a consequence, there is an urgent need to develop X-ray detectors with higher spatial resolution, higher sensitivity, lower detection limit and faster response rate to reduce the radiation dose, which is also a lingering challenge for researchers [9, 10]. To improve device performance, large amounts of works on action mechanism, material and structure have been conducted, with major breakthroughs made. Song and coworkers have constructed a $MAPbI_3$ single crystal X-ray detector, with sensitivity up to 7×10^5 $\mu C\ Gy^{-1}\ cm^{-2}$, and detection limit as low as 1.5 $nGy\ s^{-1}$ [11]. In term of spatial resolution, a 250 lp mm^{-1} record has been created by a $CsPbBr_3$-AAO scintillator detector [12].

Among various device structures, fibers could match the demand for X-ray detection inherently due to their geometric feature, large length-to-diameter ratio. Firstly, X-ray, a kind of high-energy electromagnetic wave, has strong penetrability. At the same time, a long fiber could provide adequate absorption depth in the direction of length and allow large absorption efficiency, which is critical for device sensitivity. In addition, the minor diameter makes a fiber a natural pixel and constructive to high spatial resolution. And a pixelated scintillator could suppress signal crosstalk better compared to its plane counterpart. In this chapter, fibers for X-ray detectors will be introduced from the aspects of operation mechanism, operation type, preparation method and device performance. Moreover, the remain challenges and potential solutions of fiber X-ray detectors will be discussed in the end to provide some practical insights and viewpoints.

2 Main Body Text

- Operation Mechanism

X-ray detectors were designed based on the interaction between X-ray and matter, mainly including photoelectric effect, Compton scattering and pair production etc., leading to energy transduction from photons to electrons. According to their operation mechanisms, X -ray detectors could be divided into the indirect and the direct. In this section, operation mechanisms of the both detectors would be explained in detail.

– Interactions between X-ray Photons and Matter

(1) Photoelectric Effect

When the energy of incident photon ($h\nu$) is greater than the binding energy (E_b)of a bound electron in a specific shell of an atom, the energy of the photon could be absorbed and transferred to the electron completely, leading to a free electron and an ionized atom. The free electron, named photoelectron, has a kinetic energy (E_k) calculated as the function:

$$E_k = h\nu - E_b \tag{1}$$

After the photoelectric interaction, a vacancy might be left in the atomic shell and filled by an electron from a higher shell, with the energy difference between the two shells carried off either by characteristic X-ray or another electron from a higher shell, called Auger electron. The filling of the initial vacancy might lead to more vacancies arising and finally leave the atom highly ionized.

(2) Compton Scattering

Compton scattering is an inelastic interaction between a higher-energy incident photon and an atomic electron, with a partial energy absorption. After scattering, the photon would deflect from the initial direct, and the electron would recoil at a certain angle as well. The Compton relationships are shown below:

$$\frac{h\nu\prime}{h\nu} = \frac{1}{1 + \alpha(1 - cos\theta)} \quad (2)$$

$$cot\phi = (1 + \alpha)tan\left(\frac{\theta}{2}\right) \quad (3)$$

$$E_k = h\nu - h\nu\prime \quad (4)$$

where

$h\nu$ is the energy of the incident photon;

$h\nu\prime$ is the energy of the scattered photon;

θ is the scattered photon angle;

ϕ is the scattered electron angle;

α is the dimensionless ratio $h\nu/m_0c^2$;

and E_k is the kinetic energy of the scattered electron.

(3) Pair Production

In the vicinity of the coulomb field of a nucleus of an electron, a high-energy photon interaction called pair production might take place. The photon energy could be converted into an electron–positron pair, so the energy balance could be descripted as:

$$h\nu = E_k^+ + E_k^- + 2m_0C^2. \quad (5)$$

As a result, pair production cannot take place when the photon energy is less than an energy threshold for the interaction of 1.02 MeV ($2m_0c^2$). And because of the instability of the positron in medium, the process usually leaves two annihilation photons.

– Operation Mechanism of Indirect X-Ray Detectors

Because of scintillators working as the active layer, indirect detectors are also called scintillator detectors. Alkali metal halides, rare earth compounds and perovskites are scintillators commonly used in X-ray detectors. As shown in Fig. 1, the scintillation conversion process could be divided into three sub-progresses: (i) the progress of converting incident radiation energy into large amounts of carriers; (ii) the energy transfer of the carriers; (iii) the emission progress of the luminescent ion radiates from the excited state back to the base state.

(1) Generation and Multiplication of Carriers

The atom become ionized after absorbing the energy from incident X-ray photon completely or partially, via the three main interaction mechanisms mentioned above. Wherein, photoelectric absorption is the dominant mechanism at a low energy level

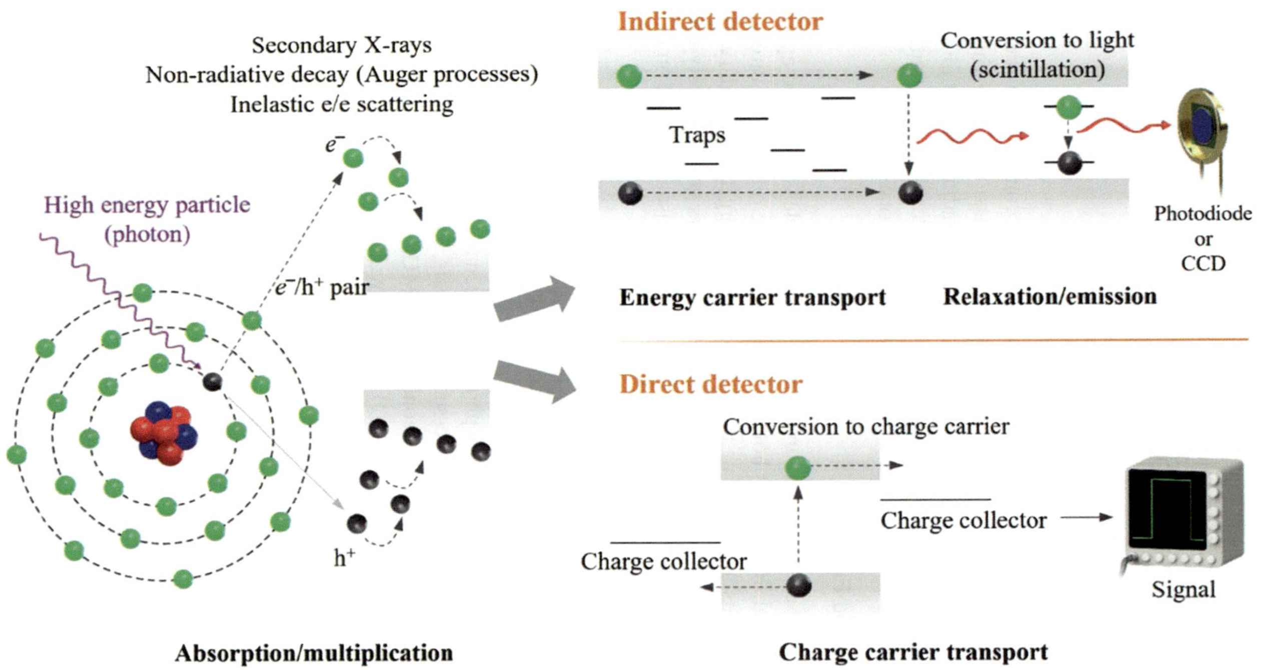

Fig. 1 Mechanism of indirect detectors and direct detectors for high-energy radiation and particles. Reproduced with permission [13]. Copyright 2022, Elsevier

(i.e., up to a few hundred keV); Compton scattering becomes dominant at energies higher than a few hundred keV; and the mechanism for pair formation only occurs above a threshold of 1.02 MeV.

The multiplication of electrons and holes is the next step, which involves various complicated processes, like secondary X-ray, non-radiative decay (Auger processes), and inelastic electron/electron scattering. At the same time, high-energy electrons and holes continue to lose energy and thermalize to the conduction band minimum and valance band maximum.

(2) Energy Carrier Transport

The radiation-generated carriers would recombine at a recombination center, rather than the site of incident radiation absorption and final emission. During the transport stage, carriers might be caught by defects, leading to nonradiative recombination and radiative recombination delays. As a result, defect state density of the material has a significant impact on the efficiency of the scintillation process.

(3) Relaxation and Emission

In this process, the energy stored in carriers would be released by two approaches: emission or thermal. Radiative photon emission could be realized by direct recombination of electrons and holes or the interaction of energetic secondary electrons with luminescent ions. Semiconductor-based scintillators generally rely on carrier recombination, while the activator-type scintillators are usually based on excitation of luminescent ions.

– Operation Mechanism of Direct X-Ray Detectors

The direct detector is also called the semiconductor type. α-Se, Si, and perovskites are commonly used semiconductor materials in direct X-ray detectors. After absorption of the energy from the incident X-ray photons, as seen in Fig. 1, the generated carriers would be directly collected and processed into signals, so a thin-film transistor (TFT) or complementary metal oxide semiconductor (CMOS) array is in need. Currently, the TFT-based flat panel detector (FPD) has been widely utilized in X-ray imaging.

- Main Classification of Fiber X-Ray Detectors

According to usage scenarios, X-ray detectors could be classified into single pixel detectors for radiation dosimeters and detector arrays for imaging.

– Single Pixel Detectors

Single pixel detectors are commonly used in the detection of radiation dosage. And detectors with fiber structures have unique advantages over other counterparts, like miniaturization, lightweight, flexibility and remote measurement. As a consequence, fibers with various internal structures have been developed to construct detectors, as shown in Fig. 2.

For indirect single pixel fiber detectors, the scintillators are usually coupled with optical fibers through different ways [14–18]. For example, a class of fiber detector tips coated with phosphor powder composite is shown in Fig. 2a. And Fig. 2b

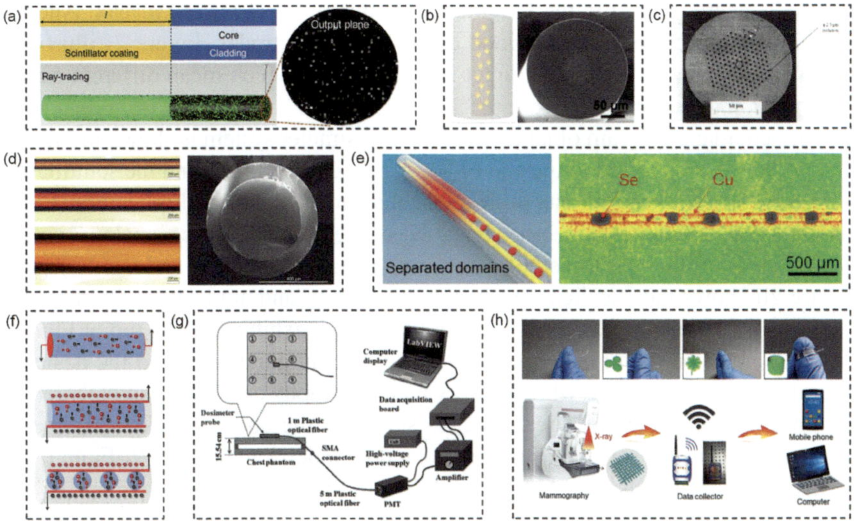

Fig. 2 Single pixel detectors with fiber structures. **a** Schematic of the fiber-optic detector with a scintillator coating, the ray-tracing process, and the cross-section of the output plane with each incident ray [19]. Copyright 2021, OSA Open Access Publishing Agreement. **b** Schematic illustration and image of the fiber detector tip with scintillator core and glass cladding. Reproduced with permission [16]. Copyright 2022, American Chemical Society. **c** SEM section images of the MOF without and with liquid scintillator filler. Reproduced with permission [14]. Copyright 2010, Optical Society of America. **d** Optical microscopy images and SEM section of direct detectors which consist of $MAPbBr_3$ perovskite fibers and silica cladding. Reproduced with permission [20] under CC BY 4.0. **e** Illustration and CT image of the Se domains and surrounding Cu electrodes in PMMA matrix. Reproduced with permission [21]. Copyright 2020, Wiley–VCH. **f** Principle illustration of direct X-ray detectors with different configurations. Reproduced with permission [21]. Copyright 2020, Wiley–VCH. **g** Schematic diagram of an indirect fiber dosimeter system. Reproduced with permission [18]. Copyright 2013, Elsevier. **h** Mechanical flexibility and application for real-time wireless monitoring of the flexible fiber shown in Fig. 2e. Reproduced with permission [21]. Copyright 2020, Wiley–VCH

illustrated a scintillator composite piece with rare earth compound core and glass cladding, which could be spiced to the distal end of an optical fiber. In addition to solid fiber, microstructure optical fiber (MOF) provides another couple mode. A liquid scintillator was utilized to fill the inclusion of the MOF, as shown in Fig. 2c. Because of the outstanding filling ability of liquid, the well-contacting solid–liquid interface with a step index would prevent the internal reflection, which might cause losses due to imperfect interface attachment.

As for direct fiber radiation dosimeters, they usually possess a core–shell structure. Figure 2d shows optical images of $MAPbBr_3$ single crystal optical fibers in glass capillaries with different diameters, and a section SEM image of a $\Phi500$ μm fiber. In addition to the continuous core, the semiconductor material could be discontinuous as well. The illustration and CT image of a flexible fiber X-ray detector are exhibited in Fig. 2e. The separated semiconductor Se domains surrounded by two Cu electrodes were inlayed in the Polymethyl methacrylate (PMMA) matrix. And the domain configuration exhibited good contact with electrodes. Figure 2f illustrates the different inner structures of direct fiber detectors, which mean different carrier transmission directions.

The schematic diagram of an indirect fiber dosimeter system is shown in Fig. 2g. The probe was connected to the end of the plastic optical fiber. And the scintillating light generated by the dosimeter probe was guided to the photo-multiplier tube (PMT) and then converted into an electric signal. Next, the output electric signal was amplified by an amplifier and transmitted to a data acquisition board, and then a laptop. Figure 2h shows the excellent mechanical flexibility and application for real-time wireless monitoring of the fiber shown in Fig. 2e. As exhibited at the top of Fig. 2h, the fiber could be rolled or woven into complex configurations. And the designed fiber device was integrated with an X-ray mammography camera to remotely record the dose rate of radiation in real time, as shown at the bottom of Fig. 2h. The X-ray-induced original signal was processed and then delivered to the Wi-Fi unit, and finally received and displayed by the mobile terminal.

– Detector Arrays

Fiber detector arrays generally do for imaging in medicine, industry and many other fields. Researches on detector arrays are currently focused on the indirect type, with its direct counterpart less explored. The scintillation layer is usually prepared via a porous template and coupled with a CCD camera or photodiode.

One of the most commonly used substrate materials in the scintillation layer is crystal silicon [24–27], which is etched to make an array of deep holes to hold scintillator, just as illustrated in Fig. 3a. And because most of the visible light photons would not propagate directly to bottom and be collected, the detection efficiency was quite low. As a result, a reflecting layer was deposited on the pore walls to make most of visible light photons could be detected. Appropriate reflective indexes of the matrix and scintillator could also ensure the light-guiding effect without reflecting layer. Generally, the refractive index of the matrix should be lower than that of the scintillator. For example, the refraction index of $Cs_3Cu_2I_5$ (the scintillator) is higher than that of AAO (anodic aluminum oxide, the matrix) in the entire visible spectrum range, as presented in Fig. 3b. Similarly, the $CsPbBr_3$ nanowires (NWs) /AAO membrane was also made to develop high-efficiency X-ray detector due to the high photoluminescence yield of metal halide perovskite. And the section SEM image in Fig. 3c exhibits the vertical alignment of $CsPbBr_3$ NWs in AAO matrix. What's more, the perovskite-based scintillator array film could also be utilized in direct X-ray detectors. Figure 3d illustrates the directed migration of the charge cloud in the $CsPbBr_2I$-AAO film. And as seen in Fig. 3a, e co-axial Si-$MAPbBr_3$ heterojunction array was also prepared for a direct panel detector, which exhibited high sensitivity and fast response speed.

Figure 3f presents a digital X-ray detector system scheme based on a pixelated scintillator. The pixelated X-ray scintillator was coupled with a fiber optic taper (FOT) array (2 × 2) to eliminate the influence of silicon pore walls which would take up part of the active imaging area of the scintillator and partially absorbed visible light.

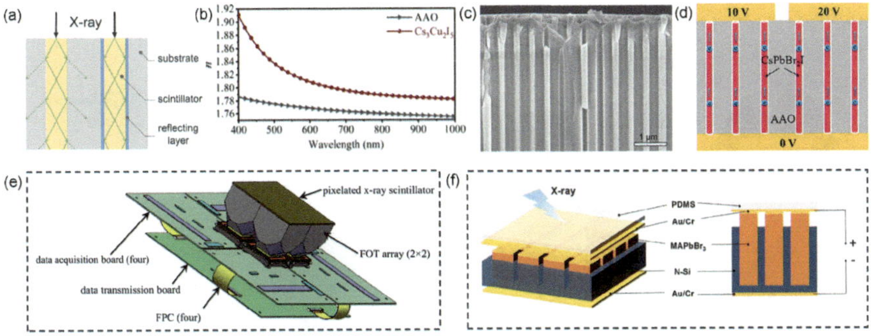

Fig. 3 Detector arrays with fiber structures. **a** Pixelated scintillator film with silicon template. **b** Refractive index of $Cs_3Cu_2I_5$ crystal and AAO. Reproduced with permission [22]. Copyright 2021, Wiley–VCH. **c** Section SEM image of the $CsPbBr_3$ NWs/AAO membrane. Reproduced with permission [23]. Copyright 2022, American Chemical Society. **d** Schematic illustration of the directed migration of the charge cloud in the $CsPbBr_2I$-AAO film. Reproduced with permission [12]. Copyright 2022, Elsevier. **e** Scheme of the co-axial Si-$MAPbBr_3$ heterojunction arrays. Reproduced with permission [24]. Copyright 2020, Elsevier. **f** Schematic representation of the digital X-ray detector system. Reproduced with permission [25]. Copyright 2015, Elsevier

The acquisition and transmission of data was performed by data acquisition boards (DABs) and a data transmission board (DTB). And the detector was connected to an upper computer.

- Preparation of Fiber X-Ray Detectors

Because single pixel detectors and detector arrays possess different structures, the preparation of fiber X-ray detectors would also be introduced from the two categories and further classified.

– Preparation of Single Pixel Detectors

(1) Drawing Method

Drawing method is one of the most widely used approaches to process fibers [16, 21]. As to single pixel detectors, drawing could be applied to prepare scintillator tips of indirect fiber detectors and fiber of direct detectors. Figure 4a schemes the thermal-drawing process of Ce/Tb-doped yttrium aluminum garnet (Ce/Tb: YAG) fiber, the scintillator fiber. First the prepared Ce/Tb: YAG crystal rod was inserted into a silica tube, then co-heated and drawn into a fiber. In addition to the simple co-axial structure, thermal-drawing could also be used to acquire fibers with complex core–shell structure. As illustrated in Fig. 4b, Se semiconductor and Cu filament were selected as the porotypes of the radiation-harvesting and charge-collecting unit, and scaled down to desired size. Significantly, heating temperature would impact the continuity of filler in the matrix during drawing. Under appropriate temperature, the continues Se core would be broken into Se domains, just as Fig. 1e shows.

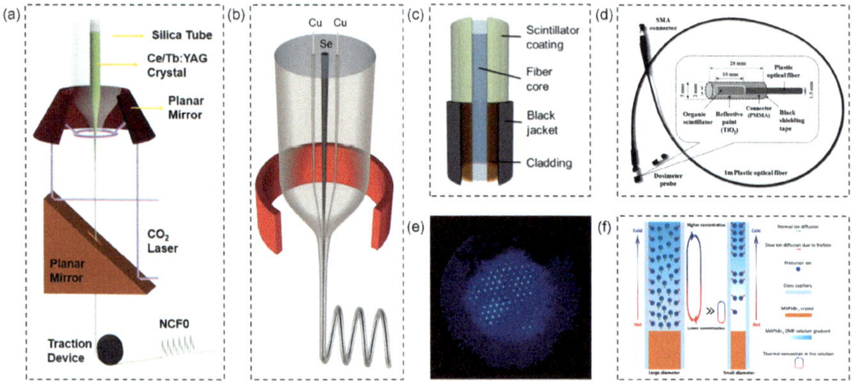

Fig. 4 Preparation of single pixel detectors. **a** Schematic of the Ce/Tb: YAG precursor fiber drawing process. Reproduced with permission [16]. Copyright 2022, American Chemical Society. **b** Illustration of the multimaterial fiber drawing process. Reproduced with permission [21]. Copyright 2020, Wiley–VCH. **c** Illustration of the optical fiber tip with scintillator coating. Reproduced with permission [19]. Copyright 2021, OSA Open Access Publishing Agreement. **d** Detailed structure of the dosimeter probe and image of the fiber detector. Reproduced with permission [18]. Copyright 2013, Elsevier. **e** Section image of MOF filled with liquid scintillator under excitation. Reproduced with permission [14]. Copyright 2010, Optical Society of America. **f** Illustration of the in-situ synthesis mechanism of $MAPbBr_3$ single crystal fiber. Reproduced with permission [20] under CC BY 4.0

(2) Coupling Scintillator with Optical Fiber

It's an important approach to construct a fiber detector by attaching scintillators onto the end of fibers. The emission under radiation could be transmitted through the optical fiber, achieving remote detection. Coating is a facile approach to combine optical fiber and scintillator with powder state. As seen in Fig. 4c, before coating with phosphor paste, the black jacket and outer cladding of the fiber end were removed firstly. Then the phosphor composite was applicated to cover the exposed fiber core tip, followed by solidification. As to a bulk scintillator piece, splicing to an optical fiber's end face is also a convenient and widely-used approach. The scintillator pieces could be either pure scintillators or composites. Figure 4d exhibits detailed structure of a fiber tip with a 10 mm long organic scintillator rod. Different from coating, the bonding between the optical fiber and the scintillator piece is significantly weak, so a connector is generally added outside. What's more, liquid scintillator could also be used in X-ray detection, filling the MOF via capillary action. As shown in Fig. 4e, the fiber filled by liquid scintillator with 425 nm emission was prepared and exhibited great optical property.

(3) In-Situ Synthesis of Radiation-Harvesting Core

In regard to direct detectors, the core–shell structure fibers could also be prepared by in-situ synthesis, other than drawing. Lead halide perovskite is one of the promising materials for X-ray detection. The synthesis mechanism of $MAPbBr_3$ single crystal fiber is illustrated in Fig. 4f, where capillaries could guarantee the axial directional growth and limit the growth in the radial direction. And the crystallization was simply controlled by temperature gradient along the capillary filled with precursor solution.

– Preparation of Detector Arrays

Detector arrays are usually prepared based on porous templates, with scintillators or semiconductors filling the pores. And the approach of filling could be mainly classified into pressure filling and in-situ synthesis.

(1) Pressure Filling

Pressure filling is a classical filling method due to its compatibility of different state of fillers and pressing temperatures. Figure 5a exhibits the schematic of the infiltration method of bismuth polymer at room temperature. The pressure differential leads to a uniform infiltration of the plastic scintillator into the porous plate. This approach is reported in 2018 [28]. Room temperature pressing could also applied to phosphor nanocrystals. The prepared pure $CsPbBr_3$ NCs was added on the AAO template, followed by negative pressure filling in a vacuum drying oven. And the diagram of preparation process of $CsPbBr_3$-AAO arrays was exhibited in Fig. 5b, with images of AAO and $CsPbBr_3$-AAO arrays below. Hot pressing is usually used in pores filling for inflexible or nonliquid scintillators, which are prepared in advance. And the solvent-free method could suppress pinholes effectively. As presented in Fig. 5c, the prepared $Cs_3Cu_2I_5$ powder was first heated on a quartz to melt. Then a piece of AAO template was placed, making the $Cs_3Cu_2I_5$ melt soak into the pores. And in order to further remove the residual air, another piece of quartz was added to provide pressure from above. Ultimately, the $Cs_3Cu_2I_5$-AAO film used in scintillator detector could be acquired with the temperature decrease. Similarly, the preparation of $CsPbBr_2I$-AAO thick film for a direct detector was exhibited in Fig. 5d. AAO template was placed on the FTO substrate and covered with $CsPbBr_2I$ powder, different from Fig. 5c. And then the mixture was calcinated in air to obtain a complete scintillator melt. Under capillarity action, the melt was squeezed into the holes by another FTO glass. Finally, a slow cooling rate was adopted to avoid residual stress.

(2) In-Situ Synthesis

In-situ synthesis could also be utilized in the preparation of detector arrays, just as shown in Fig. 6. The scheme in Fig. 6a exhibited the synthesis process of the $MAPbBr_3$ in pores of Si substrate. $PbBr_2$ was first fed into the holes and reacted with the diffuse MABr gas at 170 °C, from top to bottom over time. And finally, the $PbBr_2$ converted into $MAPbBr_3$ completely. The synthesized $MAPbBr_3$ pillar had uniform composition at different depths, as shown in Fig. 6b. Besides solid–gas reaction, the synthesis could be carried out by means of solvent as well (Fig. 6c). A certain amount of precursor solution was dropped on a glass slide, followed by an AAO membrane placed on the droplet. After the solution filling the pores completely under capillary forces, solvent evaporation proceeded at 70 °C from the top side. During the evaporation, precursor solution beneath the AAO continuously supplied the consumption, which guaranteed the continuous growth of $CsPbBr_3$ nanowires. And the selected area electron diffraction (SAED) pattern and low-magnification

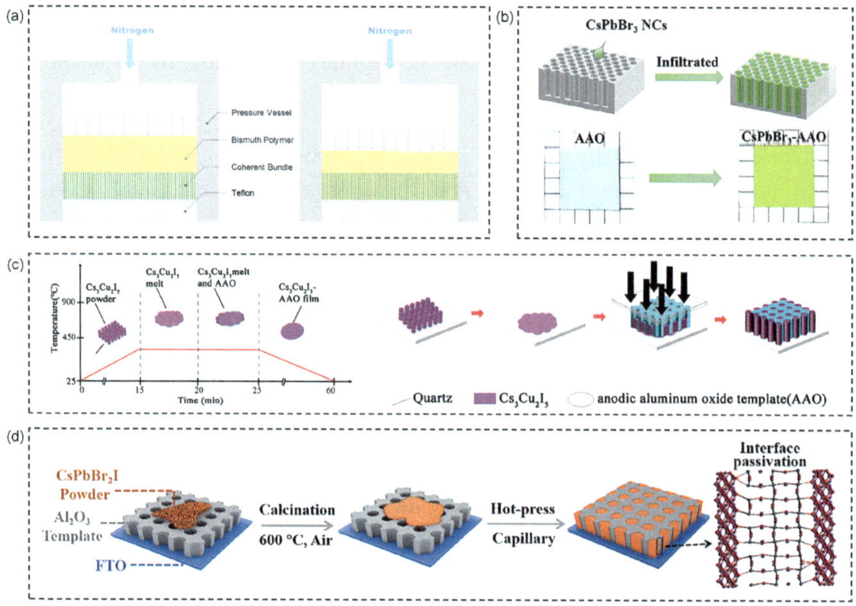

Fig. 5 Preparation of detector arrays via pressing filling. **a** The pressing scheme of polymer scintillator at room temperature. **b** Illustration of the negative pressure filling for $CsPbBr_3$ NCs into AAO template. Reproduced with permission [29]. Copyright 2021, Wiley–VCH. **c** The temperature controlling procedure and illustration for fabrication of $Cs_3Cu_2I_5$-AAO film. Reproduced with permission [22]. Copyright 2021, Wiley–VCH. **d** The schematic illustration of the $CsPbBr_2I$-AAO film via hot-pressing. Reproduced with permission [12]. Copyright 2022, Elsevier

TEM image of a single nanowire extracted from AAO pores were exhibited in Fig. 6d, proving the single crystalline. In addition, directionally solidified eutectic (DSE) system could also be applied to prepare phase-separated scintillator fibers (PSSFs). The eutectic $GdAlO_3$ (GAP [gadolinium aluminum perovskite]) -Al_2O_3 DSE ceramic doped with Ce^{3+} was prepared, with diagonal view and SEM images exhibited in Fig. 6e–g. The GAP fibers were seen well-aligned and surrounded by Al_2O_3 matrix.

- Performance of the Fiber X-Ray Detectors

This section summarizes some of the fibers used for high-energy radiation detection and imaging. For X-ray imaging, Fig. 7 shows some of the imaging photographs acquired via the fiber detector arrays mentioned above. Figure 7a shows the imaging plane illustrations, photographs and computerized tomography images of four objects made from different materials. As seen in Fig. 7a, details of the objects could be clearly reflected by the CT images. And Fig. 7 b presents the flat-field corrected projection, 3D visualization of reconstructed volume and horizontal slice through reconstructed volume of Siljan rock grain. Different parts of the sample could be distinguished clearly in the tomography. Figure 7c shows the images of a memory card taken by the $CsPbBr_3$-AAO arrays (left) and $CsPbBr_3$ film (right) scintillation

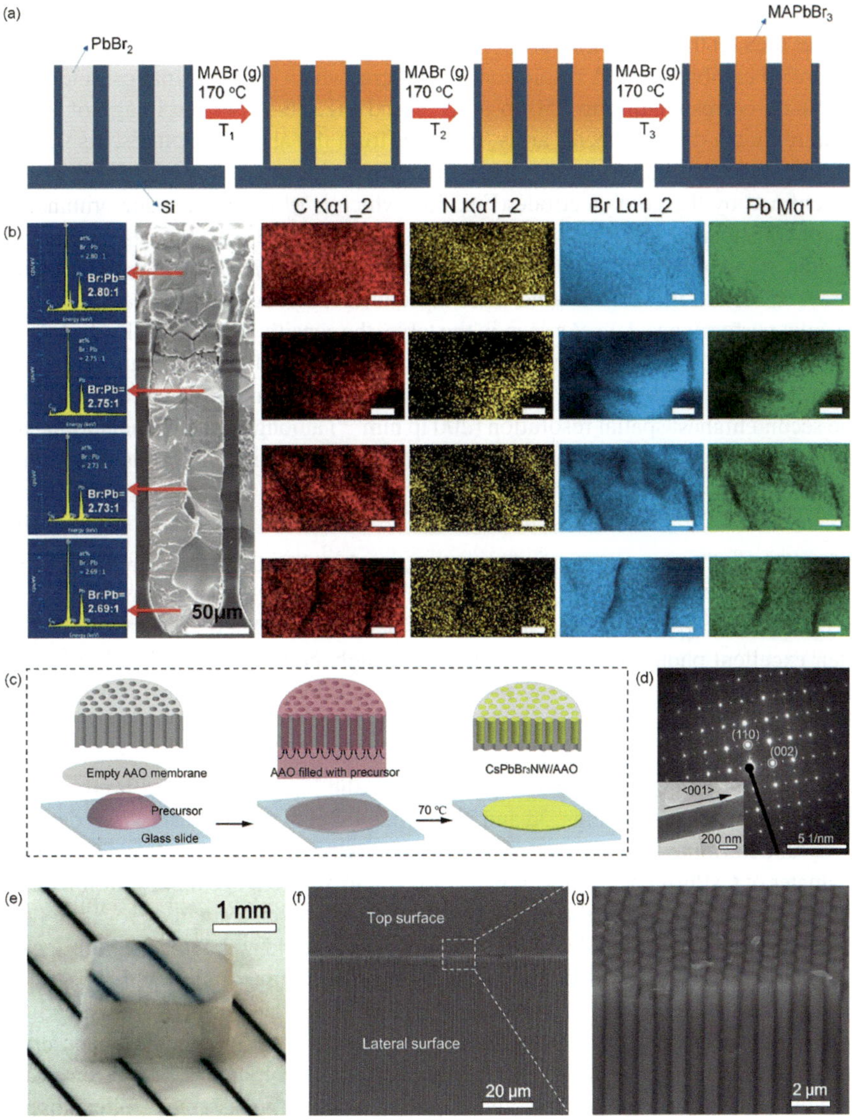

Fig. 6 Preparation of detector arrays via in-situ synthesis. **a** Schematic diagram of the forming process of the $MAPbBr_3$ perovskite in Si template. **b** EDS spectrum of the as-fabricated $MAPbBr_3$ pillar at different depths. All scale bars are 1 μm. Reproduced with permission [24]. Copyright 2020, Elsevier. **c** Schematic diagram of the growth process of $CsPbBr_3$ NWs in AAO membranes. Reproduced with permission [23] under CC BY 4.0. **d** Selected area electron diffraction (SAED) pattern. The inset shows low-magnification TEM. Reproduced with permission [23] under CC BY 4.0. Diagonal view (**e**) of the cut and polished sample, SEM image (**f**) and enlarged SEM image (**g**) of the sample edge. Reproduced with permission [30]. Copyright 2013, American Institute of Physics

screen respectively. The image on the left is much clearer than the right one, which proves the fact that fiber structure contributes to reduce spatial resolution in imaging. Figure 7d exhibits a JIMA resolution test pattern, which shows a smallest line space of 2 μm, corresponding to 250 lp mm^{-1}. And Fig. 7e shows the image of letters acquired via a direct detector array, different from the above 4 pictures. As seen in the figure, the spatial resolution of Fig. 7e is not high enough, because the pixel size is confined by the gold electrode (Fig. 3d), which is of a square shape with a side length of 200 μm.

And some of the main characteristic parameters of fiber dosimeters and imaging arrays are listed in Table 1. And there are considerable researches that present no specific performance data. As seen in the table, the sensitivity of the $CsPbBr_2I$-AAO direct detector array is up to 4581 μC Gy^{-1} cm^{-2}. And the minimum detection limit is 0.033 nGy s^{-1}, from an indirect $CsPbBr_3$-AAO array, which also exhibits the second highest spatial resolution (200 lp mm^{-1}) among the listed detectors. And another $CsPbBr_3$-AAO array possesses the highest resolution of 250 lp mm^{-1}.

- Analysis and Discussion

According to the parameters in Table 1, perovskite-based detectors exhibit better performance. As a result, it could be concluded that lead halide perovskite is one of the most competitive candidates for high-performance X-ray detectors, owing to their excellent photoelectric performance and high atomic number. In addition, the number of researches on fiber X-ray detectors of direct type is much less than that of the indirect, especially in the field of imaging.

Even though the direct type has significant advantages over the indirect, the device structure and preparation technology might be the biggest obstacles hindering its development. As seen in Figs. 3d and 7e, the pixel size of the $CsPbBr_3$-AAO detector array is determined by electrodes, not the pore size of AAO template. The minor diameter of $CsPbBr_3$ NWs unfortunately makes no difference to the resolution under the plane structure. As to the $MAPbBr_3$-Si heterojunction, a single perovskite fiber could operate as a pixel under plane structure, which could guarantee the high spatial resolution. But there is still a considerable challenge on integration with TFT.

As a result, the potential solutions might be proposed from two perspectives, improving the preparation technology of plane device or optimizing the device structure. Photoetching and transfer printing might be practicable approaches to reduce the size of electrode and enhance resolution [33]. However, the instability of lead halide perovskite should be noticed. Some reagents involved might result in the breakdown of perovskite. On the other hand, the sandwich structure of electrodes and semiconductor layer in each fiber might be replaced by co-axial structure, in which the size of single pixel could be strictly confined in the range of fiber diameter. In addition, the thickness of the semiconductor layer is controllable, which means a desired carrier transmission distance and electron collection efficiency. While the interface problem caused by curved face should be taken into consideration.

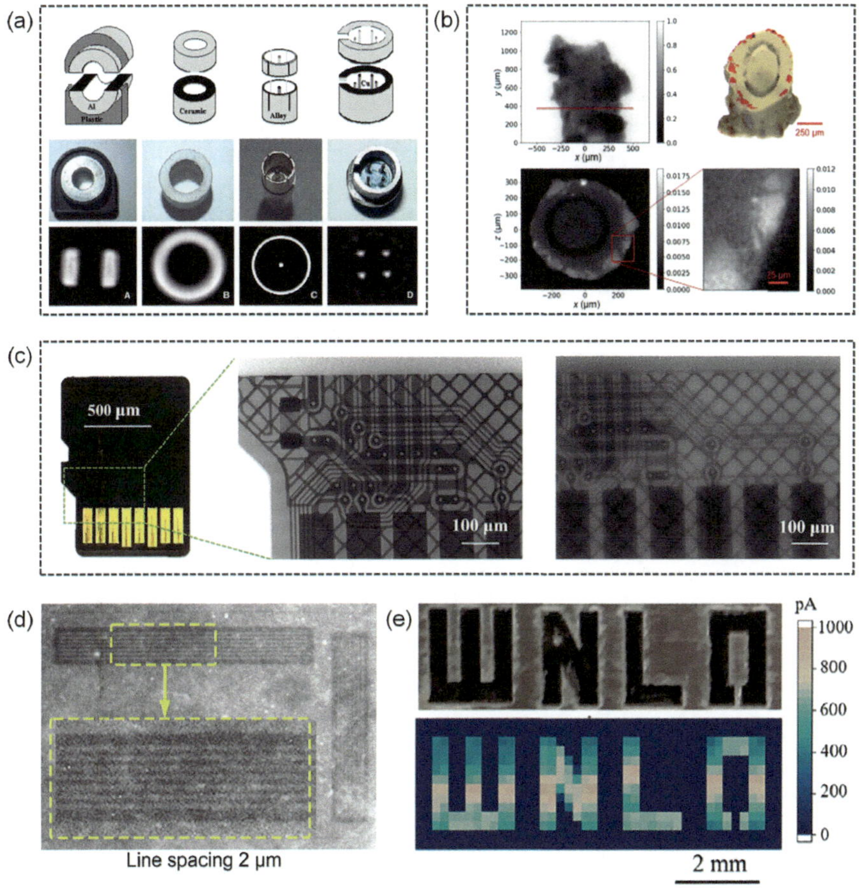

Fig. 7 Imaging performance of the fiber detector arrays. **a** Illustrations, photographs and computerized tomography images of four tested objects. Reproduced with permission [27]. Copyright 2008, American Association of Physicists in Medicine. **b** Flat-field corrected projection, 3D visualization of reconstructed volume and horizontal slice through reconstructed volume of Siljan rock grain. Reproduced with permission [31]. Copyright 2023, Springer Nature. **c** X-ray imaging of memory card (green dashed square) with the pixelated CsPbBr3-AAO arrays (left) and CsPbBr3 film (right) scintillation screen. Reproduced with permission [29]. Copyright 2021, Wiley–VCH. **d** X-ray image of a JIMA test pattern with a line spacing of 2 μm. Reproduced with permission [23] under CC BY 4.0. **e** The optical (top) and X-ray (down) images of the logo were obtained by the $CsPbBr_2I$/AAO detector. Reproduced with permission [12]. Copyright 2022, Elsevier

3 Conclusion

X-ray has been making increasingly difference to technology and life since the discovery in 1895. But with deeper insight in X-ray, the potential damage to human health is attracting more attention to the reduce of radiation dosage, where X-ray

Table 1 Characteristic parameters of fiber X-ray detectors

Materials	Operation mode	Sensitivity (μC Gy^{-1} cm^{-2})	Detection limit (nGy s^{-1})	Spatial resolution (lp mm^{-1})	References
CsI:Tl/Si	Indirect	–	–	20	[25]
Triphenolbismuth(19%)-polyvinylcarbazol	Indirect	–	–	50	[28]
Gadolinium oxysulfide	Indirect	–	~0.26	–	[19]
$Cs_3Cu_2I_5$-AAO	Indirect	–	–	10	[22]
$CsPbBr_3$-AAO	Indirect	–	0.033	200	[29]
$Cs_3Cu_2I_5$-AAO	Indirect	–	–	20	[32]
$CsPbBr_3$-AAO	Indirect	–	–	250	[23]
$CsPbBr_3$-AAO	Indirect	–	–	180 ± 20	[31]
$MAPbBr_3$-Si	Direct	242	8.5	–	[24]
$CsPbBr_2I$-AAO	Direct	4581	~136	–	[12]

detectors act an important role. According to the interactions between detector materials and X-ray, the operation mechanisms could be divided into the indirect and direct. Fiber detectors have unique advantages over other structures owing to the geometric feature. As to single pixel detectors, the fiber structure could help achieve miniaturization, lightweight, flexibility and remote measurement. And the fiber structure possesses more significant advantages in X-ray imaging, where the small diameter has a remarkable impact on the reduction of spatial resolution. In addition, the fiber probes are commonly easy to prepare, based on optical fibers or mono-porous or porous templates. Because of the excellent photoelectric characteristics, lead halide perovskites exhibit outstanding performance as both scintillator and semiconductor in previous researches. As a result, lead halide perovskites are considered as competitive candidates for next generation efficient X-ray detectors. It is obvious that indirect detectors are explored more than the direct. And the direct ones perform not as well as the indirect, which is against the theoretical forecasting. The remain challenge of direct fiber detectors might be focused on the preparation technology and device structure, where more effort should be made.

References

1. Allen HS. Nature. 1931;127:356–8.
2. Kaye GWC. Nature. 1934;133:511–3.
3. Meng LJ, Iwanczyk J. IEEE Trans Med Imaging. 2016;35:710–1.
4. Mino L, Borfecchia E, Segura-Ruiz J, Giannini C, Martinez-Criado G, Lamberti C. Rev Mod Phys. 2018;90: 025007.
5. Liu Y, Sun Q, Liu J, Norouzi Banis M, Zhao Y, Wang B, Adair K, Hu Y, Xiao Q, Zhang C, Zhang L, Lu S, Huang H, Song X, Sun X. ACS Appl Mater Interfaces. 2020;12:2293–2298.
6. Chambers CE. Circulation. 2017;136:2417–9.
7. Udroiu I, Sgura A, Chendi A, Lasagni L, Bertolini M, Fioroni F, Piccagli V, Moramarco A, Romano MG, Fontana L, D'Alessio D, Bruzzaniti V, Rosi A, Grande S, Palma A, Giliberti C, Iori M, Piergallini L, Sumini M, Isolan L, Cucchi G, Compagnone G, Strigari L. Sci Rep. 2020;10:21693.
8. de Oliveira PMC, do Carmo Santana P, de Sousa Lacerda MA, da Silva TA. Radiat Phys Chem. 2017;140:305–308.
9. Chen M, Sun L, Hong Z, Wang H, Xia Y, Liu S, Ren X, Zhang X, Chi D, Yang H, Hu W. ACS Appl Mater Interfaces. 2022;14:41275–82.
10. Wei H, Fang Y, Mulligan P, Chuirazzi W, Fang H-H, Wang C, Ecker BR, Gao Y, Loi MA, Cao L, Huang J. Nat Photonics. 2016;10:333–9.
11. Song Y, Li L, Bi W, Hao M, Kang Y, Wang A, Wang Z, Li H, Li X, Fang Y, Yang D, Dong Q. Res. 2020.
12. Zhu M, Du X, Niu G, Liu W, Pan W, Pang J, Wang W, Chen C, Xu Y, Tang J. Fundam Res. 2022;2:108–13.
13. Jana A, Cho S, Patil SA, Meena A, Jo Y, Sree VG, Park Y, Kim H, Im H, Taylor RA. Mater Today. 2022;55:110–36.
14. DeHaven SL, Albin S, Kelliher WC. Opt Express. 2010;18:13754–60.
15. DeHaven SL, Williams PA, Burke ER. AIP Conf Proc. 2016;1706: 110012.
16. Jia M, Wen J, Pan X, Zhang L, Yuan J, Huang Y, Zhang X, He L, Pang F, Wang T. ACS Appl Mater Interfaces. 2022;14:1362–72.

17. McCarthy D, Keeffe SO, Lewis E, Sporea DG, Sporea A, Tiseanu I, Woulfe P, Cronin J. IEEE Sens J. 2014;14:673–85.
18. Yoo WJ, Jeon D, Seo JK, Shin SH, Han K-T, Youn WS, Cho S, Lee B. Radiat Meas. 2013;48:29–34.
19. Chen K, Ren J, Zhao C, Liao F, Yuan D, Lei L, Zhao Y. Opt Express. 2021;29:22578–92.
20. Zhou Y, Parkes MA, Zhang J, Wang Y, Ruddlesden M, Fielding HH, Su L. Sci Adv 2022;8:eabq8629.
21. Du M, Dai Y, Wang Z, Lv S, Du G, Li J, Qiu Y, Qiu J, Zhou S. Adv Mater Technol. 2020;5:2000302.
22. Zhao X, Jin T, Gao W, Niu G, Zhu J, Song B, Luo J, Pan W, Wu H, Zhang M, He X, Fu L, Li Z, Zhao H, Tang J. Adv Opt Mater. 2021;9:2101194.
23. Zhang Z, Dierks H, Lamers N, Sun C, Nováková K, Hetherington C, Scheblykin IG, Wallentin J. ACS Appl Nano Mater. 2022;5:881–9.
24. Tian S, Sui F, He K, Cheng G, Ge Y, Ning D, Wang Z, Wang Z, Tao G, Wang Z, Du B, Wei L, Li W, Yang C, Chen M. Nano Energy. 2020;78: 105335.
25. Zhao Z, Li J, Lei Y, Wang R, Ren J, Qiao J, Niu H. Nucl Instrum Methods Phys Res A. 2015;795:71–6.
26. Guo J, Zhou B, Liu X, Lei Y, Yang Q, Niu H, Wang Y, Kuo X. Photonics Asia. 2010;2010:7845.
27. Simon M, Engel KJ, Menser B, Badel X, Linnros J. Med Phys. 2008;35:968–81.
28. Morse T, Mostovych N, Gupta R, Murphy T, Weber P, Cherepy N, Adams B, Bifano T, Stankus B, Akif A, Kingon A. SPIE Opt Eng Appl. 2018;10763.
29. Li H, Yang H, Yuan R, Sun Z, Yang Y, Zhao J, Li Q, Zhang Z. Adv Opt Mater. 2021;9:2101297.
30. Ohashi Y, Yasui N, Yokota Y, Yoshikawa A, Den T. Appl Phys Lett. 2013;102: 051907.
31. Dierks H, Zhang Z, Lamers N, Wallentin J. Nano Res. 2023;16:1084–9.
32. Hui J, Ran P, Su Y, Yang L, Xu X, Liu T, Yang YM. J Phys Chem C. 2022;126:12882–8.
33. Liu G, Tian Z, Yang Z, Xue Z, Zhang M, Hu X, Wang Y, Yang Y, Chu PK, Mei Y, Liao L, Hu W, Di Z. Nat Electron. 2022;5:275–80.

Chapter 8
Fiber Nanocomposites Based on Two-Dimensional Nanosheets

Tianzhu Zhou

Abstract The integration of two-dimensional (2D) nanosheets into fiber nanocomposites represents a cutting-edge frontier in the realms of flexible electronics. This chapter surveys recent advancements in the preparation approaches, interface construct, mechanical and electrical properties, and applications of fiber nanocomposites, emphasizing their transformative impact on electrical functionalities for flexible electronics. The incorporation of diverse 2D nanosheets, including graphene, $Ti_3C_2T_x$ (MXene), transition metal dichalcogenides, and black phosphorus, into fibrous structures has led to a paradigm shift in material engineering through the interfacial interactions. The resulting nanocomposites exhibit tailored mechanical and electronic properties, leveraging the unique attributes of 2D nanosheets to enhance light-matter interactions and facilitate efficient charge transport. These fiber nanocomposites have demonstrated versatility in the design of novel devices such as wearable energy storage, flexible sensors, and wearable energy harvesters for flexible and wearable electronic. These fiber nanocomposites boast enhanced electronic conductivity, mechanical property, and lightweight attributes, paving the way for the development of flexible electronic technologies. Looking forward, the transformative impact of these materials is anticipated to extend into the realms of human healthcare and the Internet of Things (IoT), presenting exciting opportunities for further research and technological innovation. This chapter serves as a valuable resource for researchers and practitioners seeking insights into the current state-of-the-art, challenges, and future directions in the dynamic field of fiber nanocomposites for flexible and wearable electronic applications.

T. Zhou (✉)
School of Electrical and Electronic Engineering, Nanyang Technological University, Singapore 639798, Singapore
e-mail: tianzhu.zhou@ntu.edu.sg

L. Wei (ed.), *Advanced Optical and Optoelectronic Fibers*, Advances in Optics and Optoelectronics, https://doi.org/10.1007/978-981-97-6218-7_8

1 Introduction

In the ever-evolving landscape of materials science, the integration of two-dimensional (2D) nanosheets into fiber nanocomposites has emerged as a frontier with profound implications for the advancement of flexible electronics [1]. The unique properties inherent to 2D nanosheets, such as graphene, MXene, transition metal dichalcogenides (TMDs) such, and black phosphorus [2–5], when seamlessly incorporated into fibrous structures, offer a transformative platform for engineering materials with enhanced their functionalities. The motivation behind this burgeoning field lies in the quest for materials that can simultaneously harness the exceptional properties of 2D nanosheets and the structural advantages of fibrous architectures. The characteristics of 2D nanosheets, coupled with the mechanical properties, electrical performance, flexibility, scalability, and lightweight nature of fibers, present a synergistic combination that holds promise for revolutionizing the landscape of wearable electronic applications [6].

The flexibility of fibrous structures enhances the adaptability and versatility of these electronic devices, offering avenues for applications in diverse technologies [7]. Simultaneously, the integration of 2D nanosheets into fibers has sparked significant interest in the flexible electronic device. The resulting fiber nanocomposites showcase improved mechanical flexibility, electronic conductivity, and lightweight characteristics, fostering the development of flexible and wearable electronic devices [8]. This intersection of 2D nanosheets and fibrous structures has paved the way for innovations in electronic textiles [9], flexible sensors [10], and wearable energy storage systems [11].

As embarking on this chapter, the context delves into the preparation approaches, interface construct, mechanical and electrical properties, and multifaceted applications of fiber nanocomposites for flexible and wearable electronic based on 2D nanosheets. It aims to provide a comprehensive overview of the current state of the art, addressing the challenges and opportunities that lie ahead. From the fundamental principles guiding the preparation of these materials to their applications in cutting-edge technologies, this chapter aims to illuminate the trajectory of this dynamic field. Through this exploration, we seek to contribute to the collective understanding of researchers, engineers, and scientists engaged in the exploration of fiber nanocomposites based and inspire further innovations in this exciting intersection of materials science and technology.

2 Main Body Text

- Preparation approaches

The synthesis of fiber nanocomposites, a burgeoning field in materials science, relies on a diverse array of preparation approaches to impart tailored and enhanced functionalities. The strategic combination of nanofibers with matrix materials through various techniques is pivotal in achieving superior mechanical, electrical, and structural properties, opening up new frontiers for innovative applications. In this context, different methodologies, such as thermal drawing, wet spinning, coating, electrospinning, and biscrolling [12–16], offer unique advantages and opportunities for material engineers to fine-tune the characteristics of the resulting fiber nanocomposites based on 2D nanosheets.

– Thermal drawing

The initial and inherent approach for incorporating innovative functionalities into fibers nanocomposites involves introducing new materials at specific locations during the preform stage. Subsequently, the entire assembly is thermally drawn into an elongated fiber, mirroring the fabrication process of conventional silica fibers (Fig. 1) [17], This method of thermal drawing offers the key advantage of maintaining the simplicity and scalability with several inherent attributes in the drawing process:

a. The preform preparation is prepared into a macroscopic object, utilizing straightforward and cost-effective equipment capable of controlling sub-millimeter features. Moreover, Efficient operation is imperative for achieving nanoscale structures through the plastic deformation process along the entire fiber length, typically reduces cross-section feature size [18].
b. Furthermore, through scaling down in accordance with the law of mass conservation, the volume of perform will be distributed into thin fibers nanocomposites simultaneously. This also enables the coverage of macroscopic surfaces, with instances in the fiber industry where over thousands of meters of fiber with an outer diameter of 50–1000 μm can be pulled with more functional surface area, prepared during a few hours in the thermal drawing process.
c. Fibers with polymers and 2D nanosheets exhibit high mechanical and conductive properties, making them suitable for integration within surfaces, textiles, and flexible electronic devices.
d. In addition, this preparation approach is capable of concurrently processing various micro-structures via one step for different applications.

The initial stage involves creating a macroscopic preform in the thermal drawing process based on multi-materials of different polymer and 2D nanosheets [19]. Verse materials are incorporated into the preform using different techniques to form the complex structure, for instance the rolling approach for hollow cylindrical fibers based on thin film, the consolidation approach via assembling materials with in a vacuum oven or hot press approach for solid fibers, the stack-and-draw approach, and

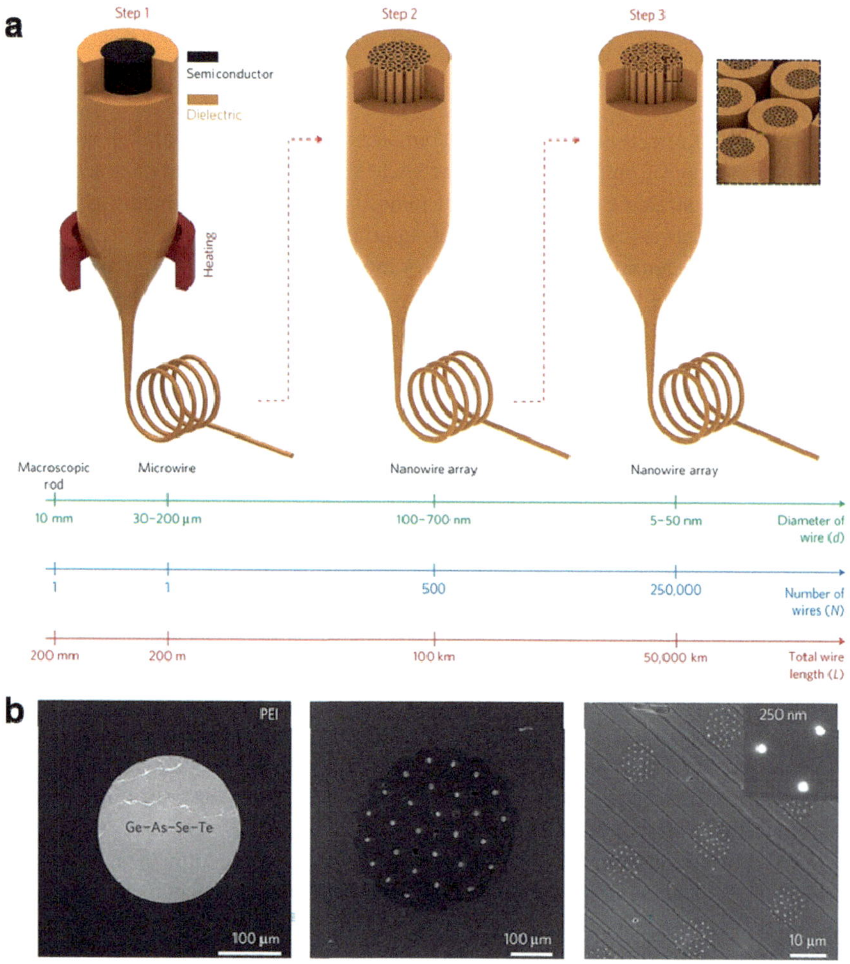

Fig. 1 **a** Successive steps of the macroscopic multi-material rod reduce to ordered arrays of nanowires by thermal size reduction with a protective matrix. **b** SEM cross-sections fibers [17]

the extrusion approach [20–24]. During thermal drawing, this resulting solid preform transforms into a thin fiber while preserving the cross-sectional microscopic structure along the entire length, contingent upon the selected thermal drawing parameters and materials.

Apart from the perform with homogeneous material with the scalability and simplicity in thermal drawing process, this approach may necessitate the simultaneous co-thermal drawing of materials with diverse thermo-mechanical properties. Due to the generation of surface area in the thermal drawing procedure, the interplay of interfacial interactions between materials becomes pivotal in determining the achievable microstructures and their geometries. In addition, the law of viscosity engineering is commonly employed to counteract this effect. Therefore, in order to constructing a multi-material fiber system, there are some parameters such as glass-transition temperature (Tg), thermal expansion, chemical reactions, crystallization, and mixing to be considered. These factors impose specific requirements, resulting in a constrained selection of multi-material combinations that can be concurrently performed within the fibers.

a. The cladding, constituting the majority of the preform and serving to maintain the assembly's cohesion, requires processing at a relatively high viscosity. This viscosity range ensures a balance: it is low enough to enable plastic deformation into fibers without fracturing, yet high enough to preserve the mechanical integrity of the desired structure. Therefore, glasses and amorphous polymers are favored choices for cladding due to their ability to be processed slightly above their Tg, maintaining the desired viscosity under an extended temperature range [18].
b. In addition to the temperature-dependent viscosity, cladding materials must be assessed for their inclination to degrade at elevated temperatures. Indeed, a significant gap between the drawing temperature and the glass transition temperature is essential in the thermal drawing process.
c. An additional concern involves the diffusion of heat between materials characterized by distinct specific heats and thermal conductivities. Given that viscosity and interfacial interactions are both material-specific and temperature-dependent, the temperature profile within the preform plays a crucial role in the drawing process. Materials with elevated thermal conductivity, such as graphene, have the tendency to act as thermal shields for other materials, and this effect must be considered.
d. It's essential to prevent a significant contrast in thermal expansion coefficients to mitigate the occurrence of excessive residual stresses in the resultant fibers, especially for various polymer cladding.
e. Materials are consolidated within a tightly confined space and elevated temperature, potentially leading to the diffusion of species and chemical interactions among different compounds. These reactions are often undesirable, although occasionally intentional and regulated.
f. Ultimately, the interplay between interfacial interactions and viscosity sets the ultimate feature sizes from the various material geometries of preforms (such as thin films, hollow channels, etc.) targeted within the fiber.

Therefore, with the thermal drawing induced stresses [19], MXene fibers nanocomposites can be prepared via thermal drawing from the hollow polycarbonate tube (Fig. 2). Via a continuous and controllable route, the ultra-compact MXene fibers were formed with in-situ generating protective layer. Consequently, the obtained MXene fibers with high orientation and low porosity not only have high tensile strength and ultra-high toughness, but also perform remarkable electrical conductivity.

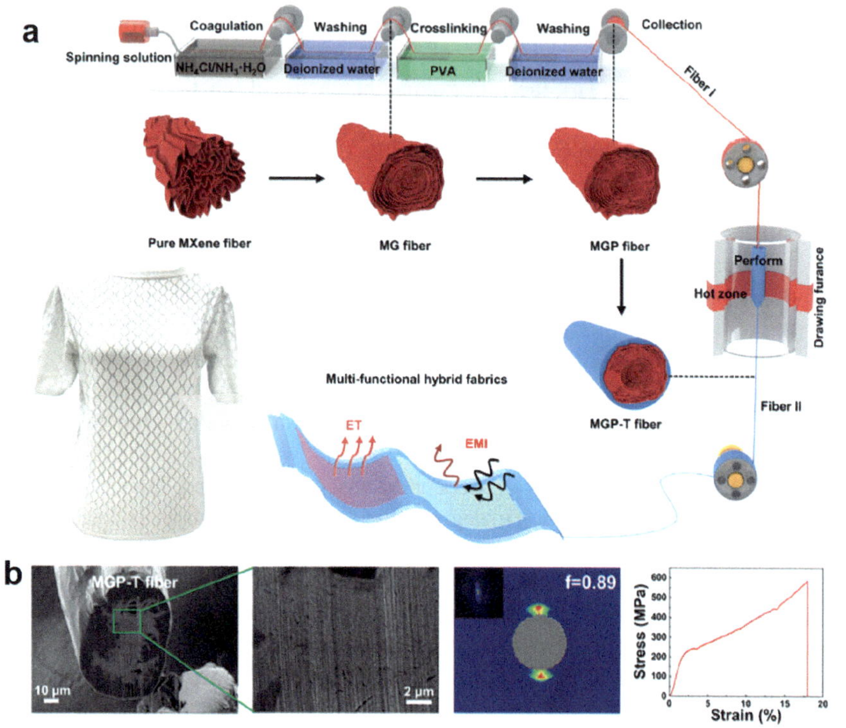

Fig. 2 **a** Preparation of MXene fiber through continuous wet spinning and thermal drawing. **b** SEM cross-sections of MXene fibers nanocomposites with polycarbonate layer and stress–strain curves [19]

– Wet spinning

Wet-spinning has emerged as a compelling approach for the fabrication of fibers nanocomposites using two-dimensional materials, offering a versatile approach to harness the unique properties of these materials in the creation of advanced fibrous structures (Fig. 3) [25]. In the context of two-dimensional materials, such as graphene oxide, MXene, transition metal dichalcogenides, and etc., the wet-spinning process involves the dispersion of these nanoscale building blocks in a liquid medium to form a stable solution or suspension. The main criteria for wet spinning approach are as follows.

a. Flake size: 2D nanomaterials, characterized by nanoscale thickness, exhibit varying lateral flake sizes, determined by synthesis methods and post-processing process. It is very important to control the lateral size, which is crucial for both the material's processability into intricate structures and the properties of the obtained fiber nanocomposites. Indeed, large flakes not only contribute to fiber

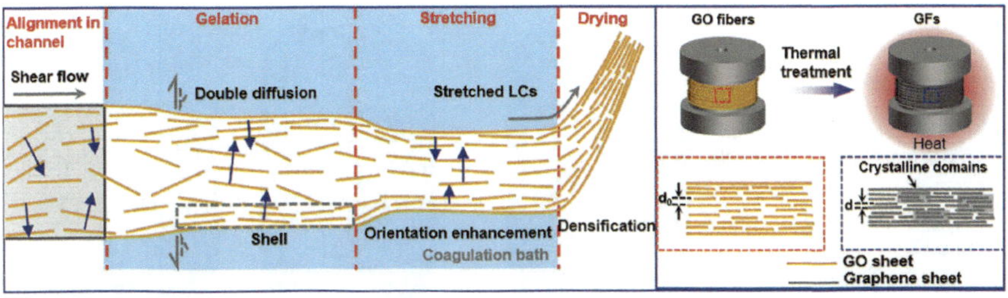

Fig. 3 The evolution schematic of graphene fibers during wet-spinning and followed by thermal treatment processes [25]

fabrication but also enhance the mechanical properties and conductivity of the resulting fibers nanocomposites. This improvement is usually attributed to a high alignment of 2D nanomaterials and a reducing of defects including voids and wrinkles. For example, using graphene oxide (GO) nanosheets with the flake size up to 100 μm and MXene nanosheets with that up to 10 μm for the nanoscale building blocks via wet spinning, their mechanical and conductive properties are significantly promoted with few of defects within fibers nanocomposites (Fig. 4) [26]

b. Solvent Selection: The choice of solvent significantly impacts the spinning of fibers. It significantly influences the viscosity of the spinning solution, stability, spinning method, and choice of coagulating bath, causing fluctuations of performance in fiber nanocomposites. Certain nanosheets, such as GO, MXene, montmorillonite, etc., can be readily processed using aqueous dispersions due to their surface terminations with water-soluble polymers (e.g., polyvinyl alcohol (PVA), cellulose, sodium alginate, etc.) [9]. Nevertheless, a variety of polymers (such as Kevlar and polyvinylidene fluoride) and other materials (such as black phosphorus and molybdenum disulfide) are not dispersible or soluble in water. To tackle this challenge, the dispersibility of 2D materials in organic solvents was investigated through uncomplicated solvent exchange procedures involving repeated centrifugation.
c. Rheological properties: In numerous fiber preparation methods, the rheological properties of the spinning solution significantly influence the capacity to produce continuous fibers or achieve uniformity. By studying the rheological of 2D materials spinning solution, the spinning process is importantly reliant on the viscosity, the loss moduli (G'), the storage moduli (G''), and the ratio of G'/G'' for the spinning solution (Fig. 5) [27]. Typically, it is necessary to ensure good flowability with a high viscosity with shear-thinning behavior when suffer from shear. Moreover, structural integrity must be guaranteed with a ratio of G'/G'' high than 1 after the removal of shear force whenever in aqueous and organic solvent (especially for GO and MXene spinning solution). In addition, the adjustment of concentration (at the range of 10–160 mg mL^{-1}) of the spinning solution allows the tuning of these parameters, enabling a broad spectrum of fabrication methods wet spinning.

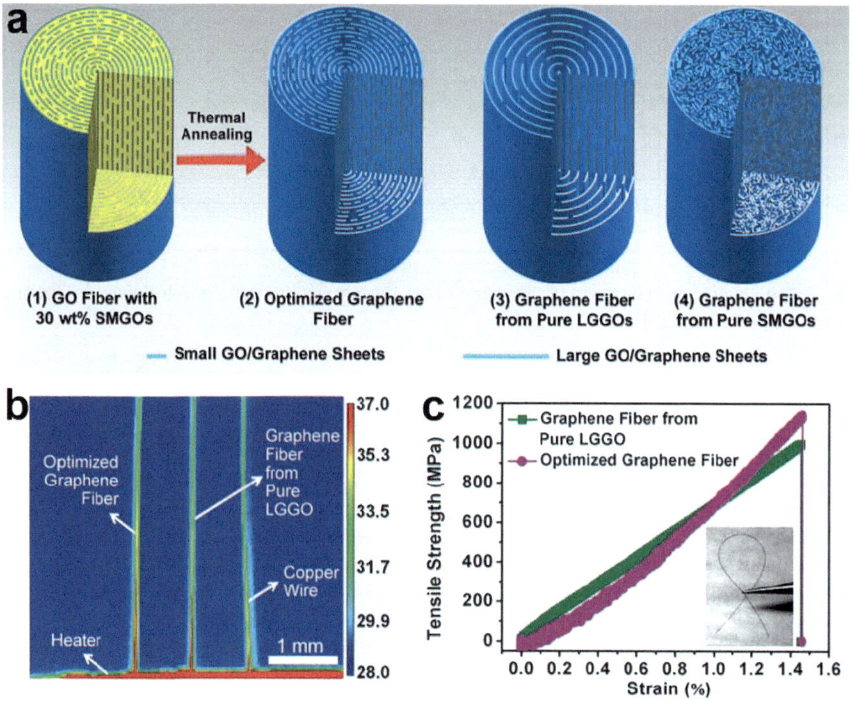

Fig. 4 **a** Filling with small GO nanosheets between large GO nanosheets. **b** A thermal image of graphene fibers. **c** Excellent mechanical strength of graphene fibers [26]

d. Coagulation bath: To initiate the wet-spinning procedure, a carefully tailored solution containing the two-dimensional materials is extruded through a spinneret into a coagulation bath. This coagulation bath serves a dual purpose: (1) it induces the rapid exchange of the solvent with a non-solvent, (2) accelerates the precipitation of the two-dimensional materials and the formation of continuous fibers. The choice of solvents and coagulation bath conditions is critical in governing the kinetics of the process and influencing the ultimate morphology and properties of the fibers [9].
e. Spinning parameters: In spinning process, its efficacy is intricately linked to the extrusion rate and collection speed. The extrusion rate, representing the speed at which the spinning solution is forced through the spinneret, is a critical parameter influencing the wet spinning process. Optimal extrusion rates are crucial for achieving uniform fiber morphology and controlling the dimensions of the resulting fibers [28]. Deviations in extrusion rates can lead to variations in fiber diameter, alignment, and overall structural integrity. Simultaneously, the collection speed, denoting the rate at which the freshly spun fibers are gathered or coagulated, plays a pivotal role in determining the final properties of the wet-spun fibers. The interplay between extrusion rate and collection speed influences the

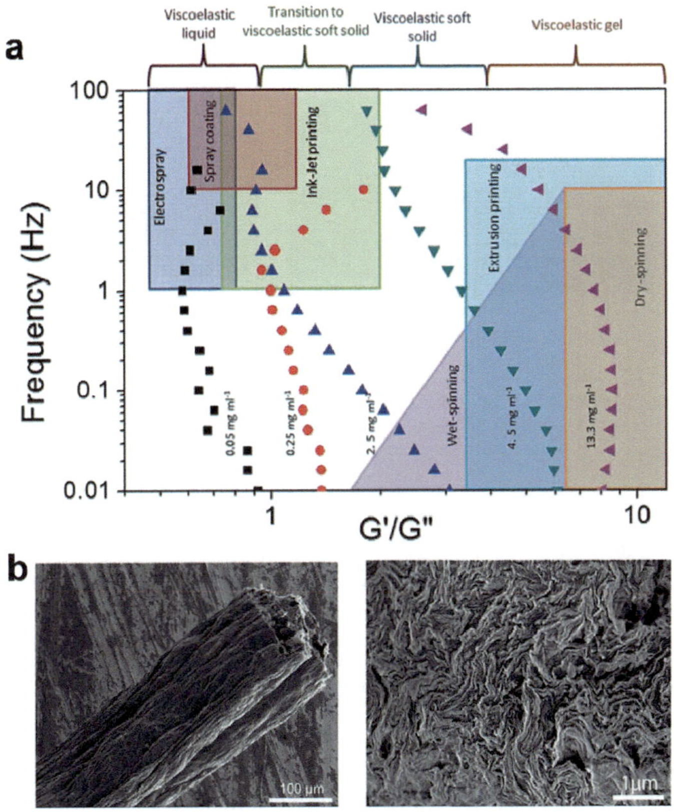

Fig. 5 **a** The ratio of loss/storage moduli of different GO concentrations conducted at a range of testing frequencies. **b** SEM images of graphene fibers [27]

stretching and alignment of the fibers during the coagulation process. Controlling the collection speed is essential for ensuring proper coagulation and preventing defects such as irregularities, buckling, or coalescence in the wet-spun fibers. A delicate balance between extrusion rate and collection speed is imperative to attain desired fiber characteristics, such as mechanical properties, conductivity, surface morphology, and porosity. Researchers continue to explore and optimize the interdependence of these parameters in wet spinning to enhance the reproducibility and scalability of the process for various applications, ranging from textiles to biomedical materials. Understanding the impact of extrusion rate and collection speed on wet spinning is fundamental for advancing the precision and efficiency of this fiber formation method (Fig. 6).

Therefore, the wet-spinning approach allows for the alignment and organization of these 2D materials during the fiber nanocomposites formation process, leading to enhanced mechanical, electrical, and thermal properties in the final product of fiber

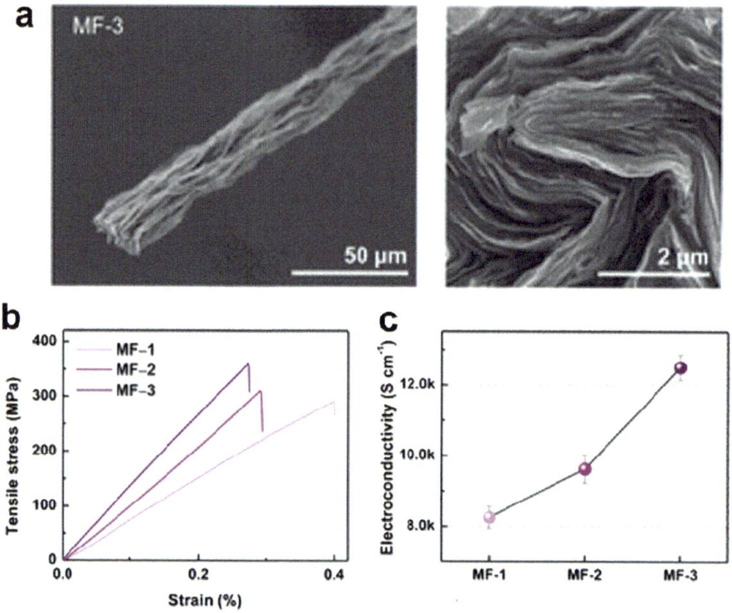

Fig. 6 **a** SEM images of graphene fibers with the increasing draw ratio. **b** Tensile strength and **c** Electroconductivity of graphene fibers with the increasing draw ratio [28]

nanocomposites. Additionally, the scalability of wet-spinning makes it an attractive technique for large-scale production of two-dimensional material-based fibers. The wet-spinning method applied to two-dimensional materials holds promise for a myriad of applications, ranging from advanced textiles to wearable electronics and reinforced composites. By precisely controlling the wet-spinning parameters, researchers can tailor the structure and performance of the fibers, opening up new possibilities for harnessing the extraordinary properties of two-dimensional materials in practical and innovative applications.

– Coating

Coating-based methods have emerged as a versatile approach for the fabrication of fibers using 2D materials, offering a unique avenue to exploit the exceptional properties of these nanomaterials in the development of advanced fibrous structures (Fig. 7). In the context of 2D materials like GO, MXene, and transition metal dichalcogenides, the coating process involves the application of a uniform layer of these nanomaterials onto a substrate due to their surface functional groups, typically in the form of a pre-existing fiber or a template that facilitates the formation of the final fibrous structure, including drop-casting, dip coating, and spray [29].

This dispersion can be in the form of a liquid solution as same as that of wet spinning. The subsequent steps involve drying or curing the coated substrate to remove any residual solvents and to promote the adhesion of the two-dimensional materials

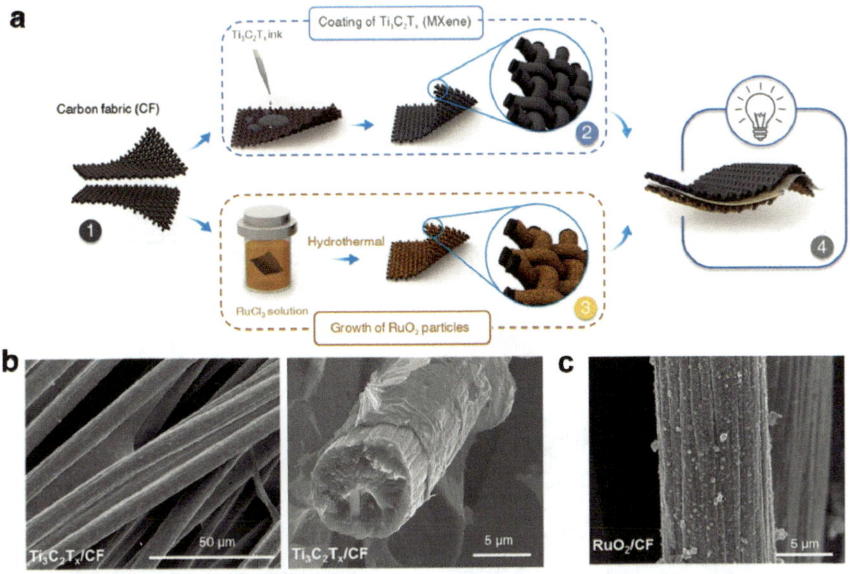

Fig. 7 **a** Illustration the fabrication process of asymmetric supercapacitor via coating. **b** and **c** SEM images of coating fibers with MXene and Ru_2O [29]

onto the substrate. Moreover, the coating process allows for a high degree of control over the thickness and distribution of the 2D materials on the substrate, enabling the creation of functionalized fibers with tailored properties. The resulting fibers exhibit enhanced mechanical strength, electrical conductivity, and other distinctive characteristics attributed to the inherent properties of the coated 2D materials.

Therefore, the versatility of coating methods makes them suitable for various applications, including the development of smart textiles and flexible electronics. By precisely adjusting the coating parameters (like flake size, solvent, rheological properties, and concentrations), researchers can optimize the structural and functional attributes of the fibers, paving the way for the integration of two-dimensional materials into a wide range of innovative and practical applications.

– Electrospinning

Electrospinning has emerged as a highly effective method for fabricating fibers using a polymer or polymer-based composite containing 2D materials, offering a versatile approach to harness the unique properties of fiber nanomaterials in the creation of advanced fibrous structures (Fig. 8). Electrospinning involves the use of an electrostatic field to draw a fine jet of polymer solution or melt containing the 2D materials from a spinneret towards a grounded collector [30].

To initiate the electrospinning process, a solution containing the dispersed 2D materials is prepared, typically involving the dissolution of polymers and the 2D nanomaterials in a suitable solvent. The solution is then electrostatically charged by applying a high voltage to the spinneret. The repulsion of charges within the solution

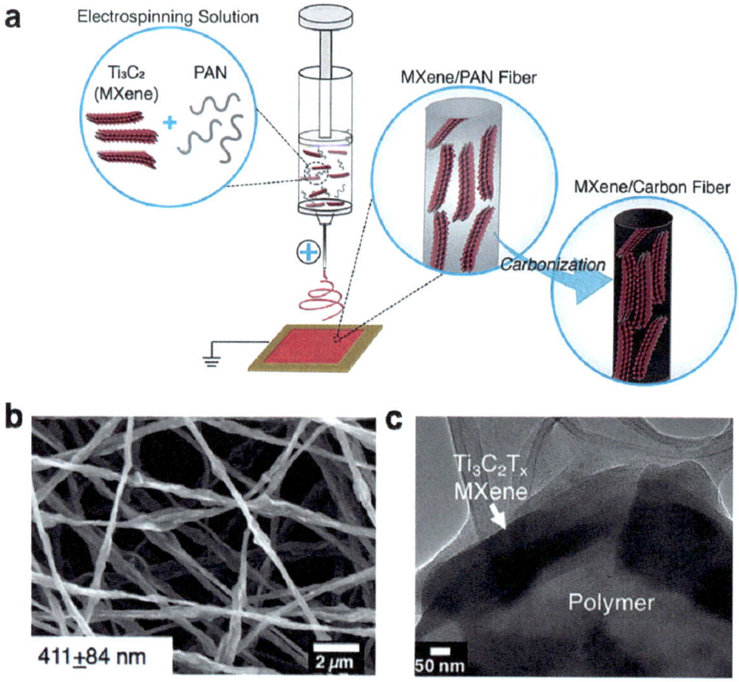

Fig. 8 **a** Illustration of the preparation of MXene/carbon nanofibers through electrospinning from MXene/polyacrylonitrile solutions. **b** SEM and **c** TEM images of MXene fibers fabricated via electrospinning [30]

leads to the formation of a fine jet, which is elongated and eventually solidifies into continuous fibers during its trajectory to the grounded collector. The collector can be a rotating drum, a flat plate, or a conductive surface, and it plays a crucial role in the deposition and alignment of the electrospun fibers.

The electrospinning method allows for precise control over the morphology within the fibers. The high surface area and aspect ratio of these materials contribute to the formation of nanostructured fibers with unique properties, such as enhanced mechanical strength, electrical conductivity, and surface reactivity. This technique is particularly advantageous for the fabrication of short fibers nanocomposites, where the 2D materials are homogeneously dispersed within a polymeric matrix. The resulting fibers find applications in diverse fields, including sensors, energy storage devices, and advanced textiles, showcasing the potential of electrospinning as a powerful tool for incorporating the extraordinary properties of 2D materials into functional and tailored fibrous materials.

– Biscrolling

Biscrolling, a novel technique in fiber preparation, involves the controlled rolling of 2D materials into fiber structures, resulting in unique and functional fibers. This method is particularly notable for its versatility in utilizing a diverse range of materials, including graphene, carbon nanotubes (CNT), GO, and MXene [9, 31]. The core concept of biscrolling lies in the controlled interaction between layers, leading to the formation of seamless and well-defined fibers. This technique offers precise control over the diameter, length, and morphology of the resulting fibers nanocomposites, making it an attractive approach for tailored applications in electronics, sensors, and composite materials. The inherent flexibility of biscrolled fibers is a notable advantage, allowing them to conform to various shapes and surfaces. One of the key advantages of biscrolling is its ability to incorporate different functional elements during the fabrication process (Fig. 9). This includes the integration of nanoparticles, polymers, or other materials, enhancing the overall properties of the resulting fibers. Additionally, biscrolled fibers exhibit intriguing mechanical, electrical, and thermal properties, making them promising candidates for advanced materials and device applications.

- Interface construct

The preparation of fiber nanocomposites based on 2D nanosheets with abundant oxygen-containing functional groups [32, 33] through interface interactions represents a sophisticated approach to tailoring material properties for enhanced performance. By harnessing various interfacial forces such as covalent bond, ionic bond, $\pi-\pi$ conjugated bond, and hydrogen bond [34–37], the resulting composites exhibit remarkable characteristics. The strategic utilization of these interactions facilitates the creation of materials with superior mechanical strength, improved electrical properties, and enhanced structural integrity. Moreover, the incorporation of multiple

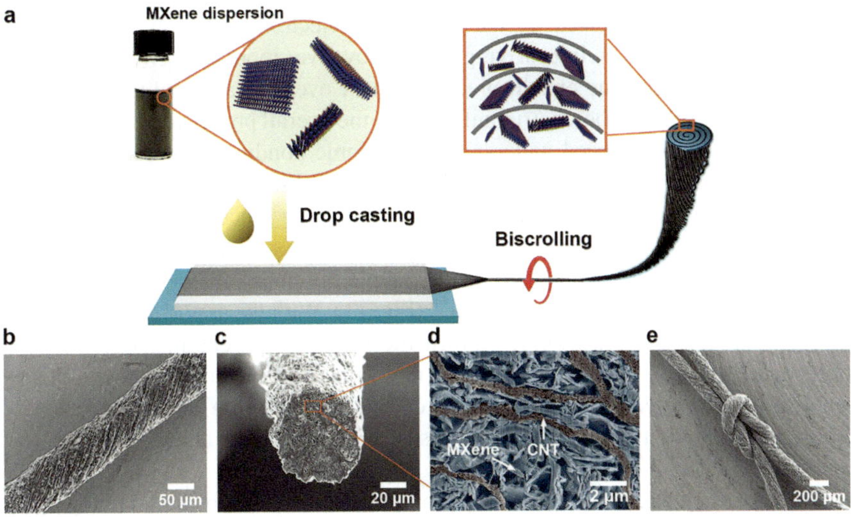

Fig. 9 **a** Illustration of the preparation of MXene/CNT fibers through biscrolling. **b–d** SEM images of MXene fibers fabricated via electrospinning biscrolling. **e** The reef knot of MXene fibers [31]

interfacial interactions synergistically contributes to the overall high performance of the fiber nanocomposites. This approach not only diversifies the avenues for material design but also opens up new possibilities for applications in various fields, ranging from structural engineering to advanced electronic devices.

- Hydrogen bond

The design and fabrication of fiber nanocomposites have witnessed significant advancements, with the utilization of interface interactions emerging as a key paradigm. Hydrogen bonding, a ubiquitous molecular interaction, are easily designed between 2D materials and polymer with the abundant oxygen-containing functional groups [32, 33]. The specificity and strength of hydrogen bonds enable controlled assembly at the nanoscale, thereby influencing the overall properties of the resulting fiber nanocomposites. A nuanced understanding of the principles governing hydrogen bonding is essential for tailoring these interactions to achieve desired material characteristics. Various fabrication techniques have been employed to exploit hydrogen bonding at interfaces, Thermal drawing (Fig. 10) [38], wet spinning, coating, and electrospinning, solution blending represent prominent methodologies for integrating nanoparticles, polymers, and other additives into the fiber matrix. Each technique offers unique advantages in terms of scalability, precision, and versatility, contributing to the diversity of applications in the field. Hydrogen bonding interface interactions exert a profound influence on the mechanical, thermal, and electrical properties of fiber nanocomposites. Through careful control of bonding dynamics, researchers can achieve enhanced structural integrity, thermal stability, and electrical conductivity, tailoring the material for specific applications.

- Ionic bond

Ionic bonding, characterized by the transfer of electrons between atoms, serves as the foundational principle driving this innovative approach. The electrostatic interactions between ions enable precise control over the assembly of nanoscale constituents within the fiber matrix, influencing the overall properties of the resulting nanocomposites. A nuanced understanding of ionic bonding principles is imperative for tailoring these interactions to achieve desired material characteristics. For 2D materials with abundant negative oxygen-containing functional groups, the fiber nanocomposites could be easily prepared with metal ions (such as Ca^{2+}, Mg^{2+}, Zn^{2+}, Fe^{2+}, and etc.) via ionic bond [25]. For example, graphene fibers can be prepared through ionic bond between GO and Ca^{2+} (Fig. 11). With the assistance of bond, the graphene nanocomposites not only perform the high tensile strength, but also have the high conductivity [37]. Moreover, another new kind of MXene fiber nanocomposites can also be fabricated via the ionic bond of NH^{4+} with high Young's modulus and electrical conductivity by reduce the porosity and enhance the alignment of MXene nanosheets [28].

Therfore, the ionic bonding interface interactions exert a substantial influence on the mechanical, thermal, and electrical properties of fiber nanocomposites. By manipulating the strength and specificity of ionic bonds, researchers can achieve tailored

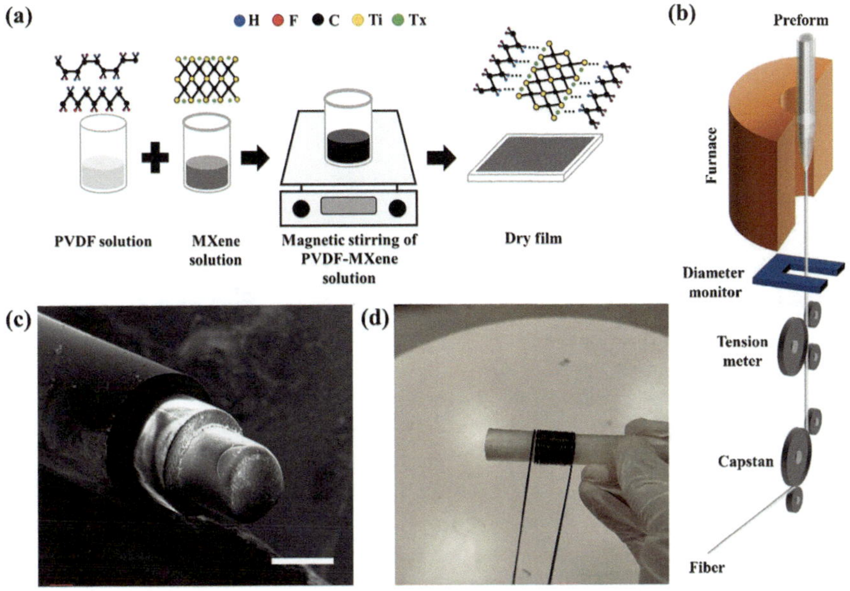

Fig. 10 **a, b** Schematic of the preparation MXene fibers through hydrogen bond between MXene nanosheets and polyvinylidene fluoride by thermal drawing. **c** SEM images of MXene fibers. **d** Photo of the flexible MXene fiber nanocomposites [38]

improvements in structural integrity, thermal stability, and electrical conductivity, allowing for customization based on specific application requirements.

– π–π conjugated bond

Because the existence of the sp^3 structure in 2D materials (such as graphene) and some organic molecules (such as pyrene derivatives, purine and pyrimidine), the π–π conjugated bond are usually constructed to prepare fibers nanocomposites like graphene fibers [36]. For example, when the Nhydroxysuccinimide ester-1-aminopyrene was introduced into the layers of graphene nanosheets, the bioinspired graphene-based nanocomposite fibers were fabricated via π–π conjugated bond through wet spinning [36]. Due to the construct of interfacial interactions, the resulting graphene fibers have a super toughness of 18.7 MJ m^{-3} with the high tensile strength of 740.1 MPa (Fig. 12). Moreover, the electrical conductivity of obtained graphene fibers is up to about 384 S cm^{-1} promoted by the π–π conjugated bond. Hence, the introduction of π–π conjugated bond do not only enhance the mechanical property of fibers nanocomposites, but also develop their conductivities.

– Covalent bond

Covalent bond is usually formed between 2D materials with the abundant oxygen-containing functional groups and some organic derivatives with -OH and -CHO functional groups or N,N′-carbonyldiimidazole [37]. Recently, the α-cyclodextrin

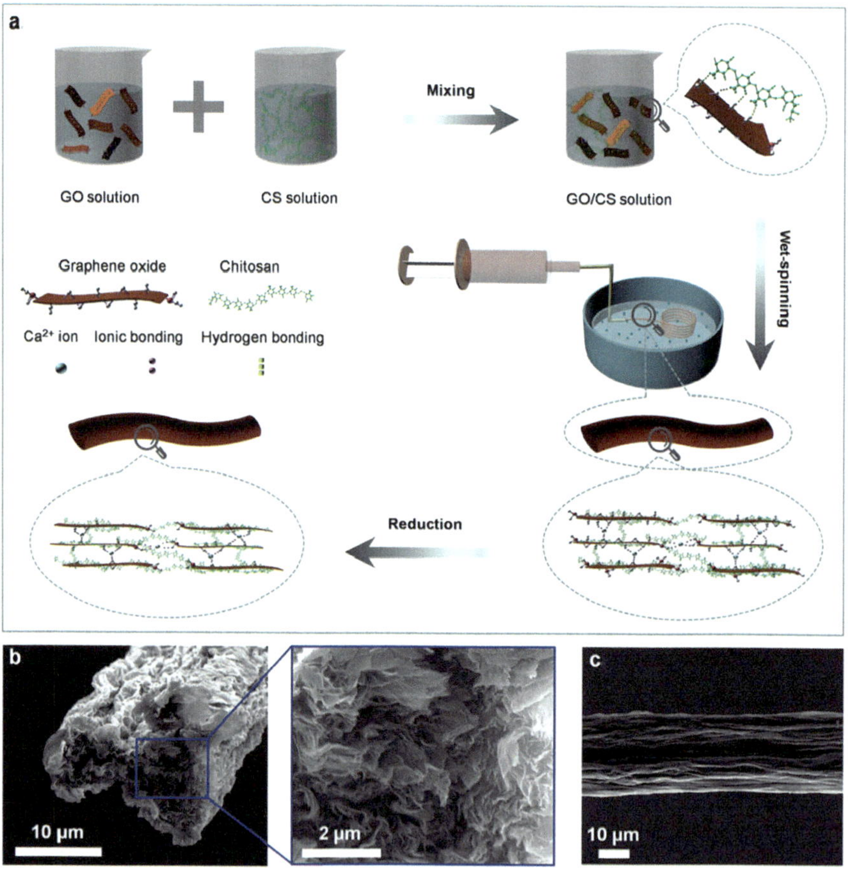

Fig. 11 **a** Schematic process of the preparation of graphene fibers through Ca^{2+} ionic bond via wet spinning. **b** SEM images of graphene fibers. **c** Surface of the graphene fiber [37]

rings of polyrotaxane was attempt to crosslink with MXene nanosheets to fabricate the MXene fiber nanocomposites via covalent bond with the assistance of N, N′-carbonyldiimidazole by wet spinning (Fig. 13) [34]. As a result, the obtained MXene fiber show the high toughness of 60 MJ m^{-3} and the high electrical conductivity of ~1100 S cm^{-1}. Furthermore, the fiber-based supercapacitor exhibited the high fiber-based supercapacitor with the capacity of ~360 F cm^{-3}, as well as 10,000 cycles under strain loading. Another instance like graphene fiber nanocomposites were also prepared through the covalent bond between GO nanosheets and 10,12-pentacosadiyn-1-ol [39]. Consequently, the resulting graphene fibers reach the high tensile strength and toughness of 842.6 MPa and 15.8 MJ m^{-3}, including the high electrical conductivity of 292.4 S cm^{-1}. Therefore, the covalent bond plays a crucial role in profoundly impacting the mechanical properties of fiber nanocomposites based on 2D materilas.

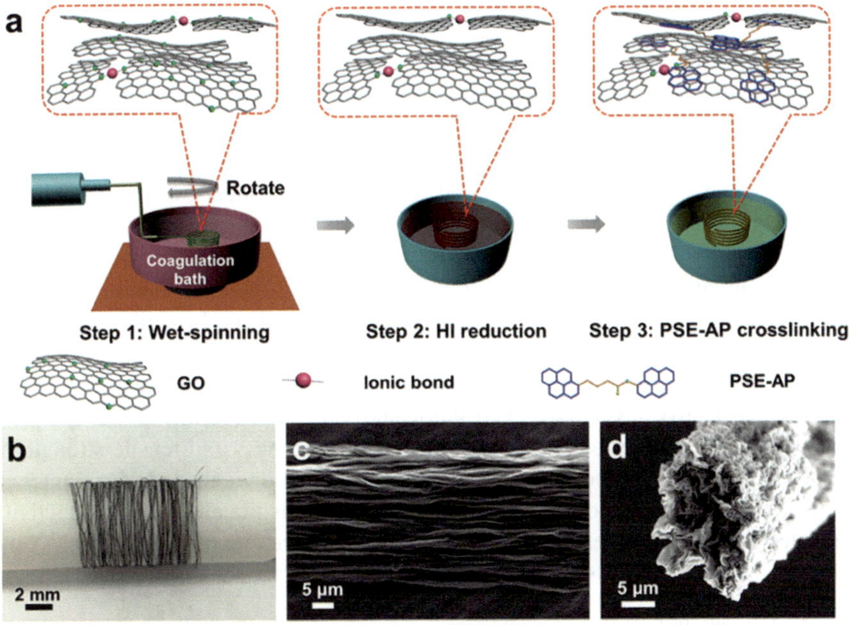

Fig. 12 **a** Schematic process of the preparation of graphene fibers through Ca^{2+} ionic bond via wet spinning. **b** SEM images of graphene fibers. **c** Surface of the graphene fiber [36]

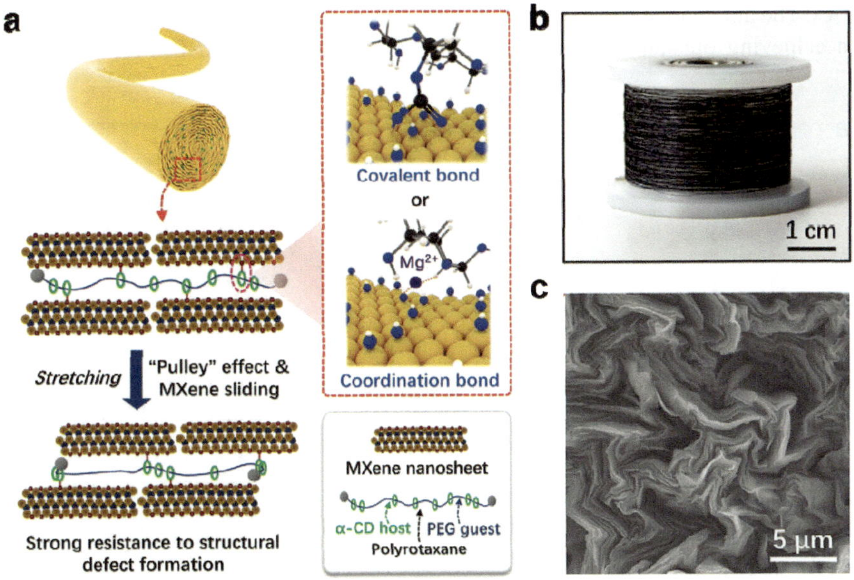

Fig. 13 **a** Schematic illustration of the preparation of MXene fibers through covalent bond via wet spinning. **b** Photo of the reel with MXene fibers. **c** SEM of the cross-section of the MXene fiber [34]

– Synergistic interfacial bond

In addition to individual interfacial interactions like covalent bonds, ionic bonds, π–π conjugated bonds, and hydrogen bonds, fiber nanocomposites can be enhanced by incorporating two or more of these interactions, leading to a superior level of performance. For example, the ultra-tough graphene fibers were prepared via the interfacial interactions of hydrogen and ionic bond among GO nanosheets, Ca^{2+}, and chitosan via wet spinning [37]. As a result, the fibers have a high toughness (26.3 MJ m^{-3}) and tensile strength (743.6 MPa) by optimizing enhancement of hydrogen bondand ionic bond. Simultaneously, the resulting graphene fiber exhibits an outstanding electrical conductivity, reaching 179.0 S cm^{-1}. This multifunctional graphene fiber holds great potential for diverse applications, including flexible supercapacitor electrodes, wearable electronic textiles, and so on. In addition, the ultra-tough MXene fibers with polymer protective layer have been fabricated via thermal drawing via hydrogen bonds and tablets interlocks (Fig. 14) [40]. The toughness and tensile strength of resulting fiber nanocomposites are significantly enhanced via the interfacial interactions with high orientation and low porosity, as well as the electrical conductivity and electromagnetic shielding performance of the woven textiles based on MXene fibers.

- Mechanical and electrical properties

The mechanical and electrical properties of fiber nanocomposites based 2D nanosheets play a pivotal role in determining their applicability across various industries. The amalgamation of nanofibers with matrix materials has proven instrumental in achieving outstanding mechanical strength, toughness, and flexibility in these

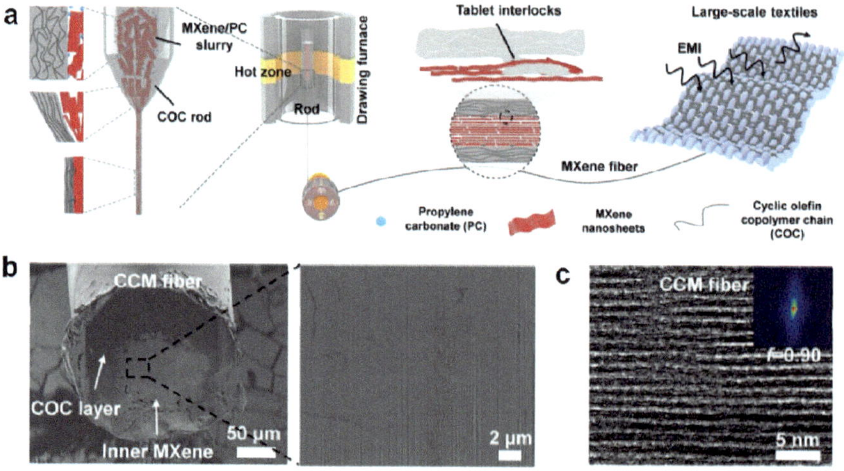

Fig. 14 **a** Schematic illustration of the preparation of MXene fibers with protective polymer layer with hydrogen bond and tablet interlocks via thermal drawing. **b** SEM image of MXene fibers. **c** TEM image of the inner MXene fiber [40]

nanocomposites. Through meticulous engineering of interfaces, such as covalent bond, ionic bond, hydrogen bond, and synergistic interfacial bond via various preparation approaches, the resulting fiber nanocomposites exhibit enhanced mechanical performance, showcasing remarkable tensile strength and toughness. Simultaneously, the incorporation of nanosheets, often composed of materials like graphene or MXene, contributes to exceptional electrical conductivity. The unique electrical properties of these nanocomposites make them highly suitable for applications in electronic textiles, conductive films, and flexible electronic devices. The synergistic interplay between mechanical robustness and electrical conductivity positions fiber nanocomposites as versatile materials with promising potential in a myriad of fields, from structural engineering to advanced electronics. This delicate balance of mechanical and electrical characteristics underscores the significance of fiber nanocomposites in addressing contemporary challenges and driving innovation across diverse industries.

- Mechanical property

The mechanical properties of individual 2D nanosheets far surpass those of the corresponding fiber nanocomposites derived from these nanosheets. For instance, graphene nanosheets possess an intrinsic Young's modulus of 1.1 TPa and a tensile strength of 130 GPa [2]. Nevertheless, the fiber nanocomposites derived from graphene exhibit subpar tensile strength and Young's modulus. Additionally, single MXene layer nanosheets boast a Young's modulus of approximately 330 GPa and a tensile strength of around 17.3 GPa [41]. Similar to graphene fibers, the current MXene fiber nanocomposites display undesirable tensile strength and Young's modulus. It is because that the mechanical characteristics of fibers nanocomposites vary widely and are contingent on factors such as interface, structure, and fabrication approaches. To augment the mechanical properties of fiber nanocomposites derived from 2D nanosheets, it is crucial to introduce interfacial interactions through diverse preparation approaches. For example, with the hydrogen bonds, MXene-coated cotton yarns demonstrated elevated tensile strength of 468 MPa [42]. It is attributed to the inherent strength of the fiber substrate with a tensile strength of approximately 334 MPa, complemented by the reinforcing effect of MXene with weight percentage of 79% (Fig. 15).

However, fiber nanocomposites composite fibers have showcased enhanced tensile strength, a result of the intrinsic mechanical properties of the host nanosheets and the interfacial interactions between the additives and nanosheets. For instance, pure MXene fibers always have the poor tensile strength from 10 to 100 MPa with different flake sizes via various spinning coagulations [43]. However, when Mg^{2+} was introduced into the interlayers via ionic bonds, the resulting MXene fibers performed the high tensile strength of 118 MPa with, much higher than pure MXene fibers [35]. In addition, the synergetic crosslinking between adjacent graphene nanosheets can be reinforced by incorporating polyvalent cations and covalent bond to connect oxygen groups or reduced GO nanosheets. For instance, the addition of Ca^{2+} and 10,12-pentacosadiyn-1-ol enhanced the mechanical strength of graphene fibers up to 842.6 MPa (Fig. 16), as well as promoting the toughness up to 15.8 MJ m^{-3} [39].

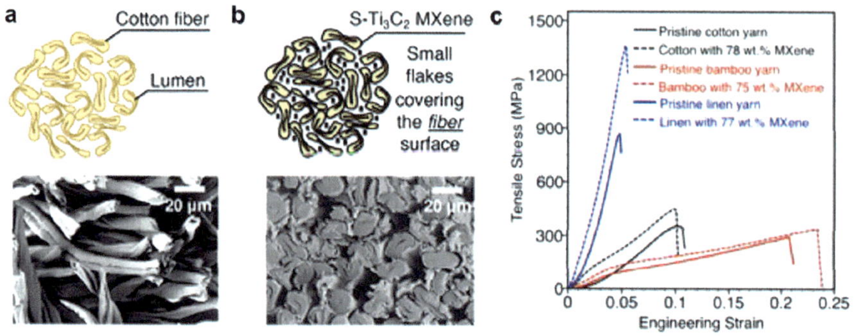

Fig. 15 **a** Carton of the cotton fiber and its SEM image. **b** Carton of the cotton/MXene fiber via coting and its SEM image. **c** Tensile stress curves of resulting fiber nanocomposites [42]

At the same time, with the crosslinking with MXene nanosheets by the interfacial of covalent bond and coordination bond, the MXene fiber nanocomposites show the high toughness of 60 MJ m^{-3} and remarkable tensile strength of ~502 MPa [34]

Moreover, the orientation of nanosheets and the porosity of the structure in fiber materials significantly influence their mechanical properties. The orientation of nanosheets within the fiber matrix determines the load-bearing capabilities and overall structural integrity of the composite material. Well-aligned

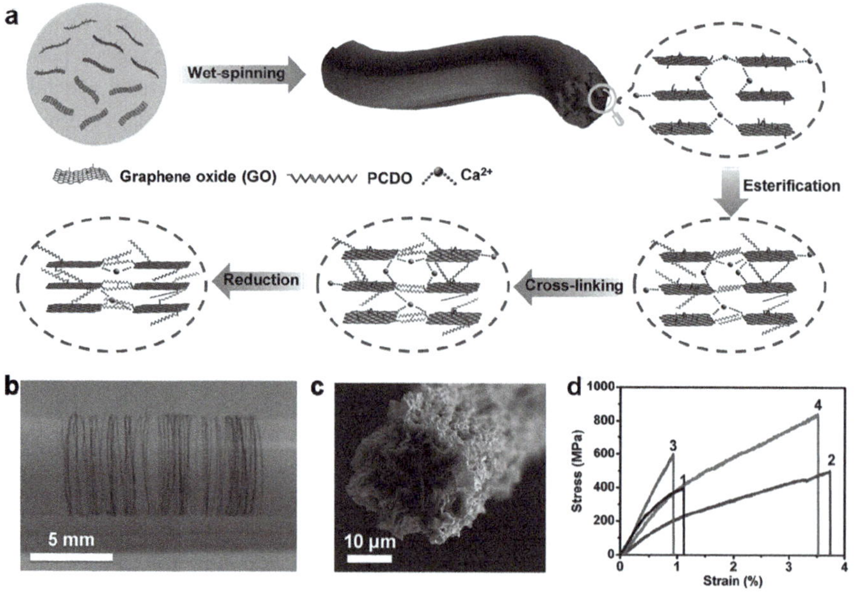

Fig. 16 **a** Illustration of the preparation of graphene fibers via ionic bond and covalent bond through wet spinning. **b** The reel with graphene fiber. **c** SEM image of resulting fiber nanocomposites. (4) Tensile strength curves of the obtained graphene fibers [39]

nanosheets contribute to enhanced tensile strength and Young's modulus along the preferred direction of alignment, reinforcing the fiber material. Conversely, random or misaligned orientations may result in less effective load transfer and weaker mechanical performance. Additionally, porosity introduces voids or empty spaces within the material, impacting its density and mechanical strength. Higher porosity generally leads to reduced density and, consequently, weaken mechanical properties, including tensile strength, toughness, and modulus. Therefore, it is important to promote the alignment of nanosheets in the fiber nanocomposite and reduce the porosity to achieve the high mechanical properties.

For example, controlling the microfluidics speed and the spinning channel (Fig. 17), the graphene fibers were prepared with the high orientation with the high tensile strength of 1.9 GPa, including the modulus of ~309 GPa [44]. Meanwhile, the aligiment of MXene nanosheets of MXene fibers are indeed promoted with the increasing drawing ratio (drawing speed: extrude speed) from 1 to3 through wet spinning, the tensile strength increase from ~304 MPa to ~343 MPa, as well as the increasing Young's modulus up to ~122 GPa. Simultaneously, the porosity of MXene fibers are decreased from 21.59% to 6.13% [28]. Furthermore, when the interfacial interactions are introduced into the interlayers of 2D nanosheets, the orientation is significantly improved with the low porosity. The resulting MXene fibers with the compact structure show the ultra-toughness of ~66.7 MJ m^{-3} and excellent tensile strength of ~585.5 MPa [19].

- Electrical property

Because of the interfacial interactions between 2D nanosheets, high orientation, and low porosity, fiber nanocomposites exhibit excellent electrical property along the axil direction. Graphene, with the sp^3 structure, perform the higher electrical conductivity via π–π conjugated bond, whenever crosslinked with the sp^3 structure

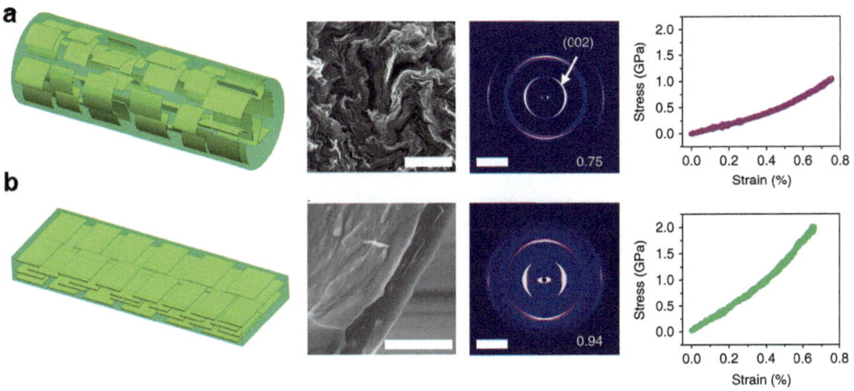

Fig. 17 Controlling the dimensions and configurations influences the orientation of sheets in solution, consequently impacting the microstructure and mechanical characteristics of annealed graphene structures: **a** Tubular channel. **b** Flat channel [44]

of some small organic moleculars [36]. By the filling of small graphene nanosheets (30 wt%) in large graphene nanosheets, the electrical conductivities of graphene fibers can reach up to ~2.2×10^3 S cm^{-1} improved by π–π conjugated bond [26]. Moreover, with the strategy of synergetic defect engineering, the graphene fibers can achieve as high as ~8.0×10^3 S cm^{-1} after high-temperature graphitization followed by the π–π conjugated bond between graphene nanosheets (Fig. 18) [45]. Apart from the graphene nanosheets, MXene nanosheets also show the notable electrical conductivity. The resulting MXene fibers can reach their conductivity up to ~7700 S cm^{-1} via wet spinning [43]. In addition, induced by the thermal drawing stresses, the MXene fibers also show the superior electrical conductivity (~1.2×10^4 S cm^{-1}) via thermal drawing [40]. Therefore, the fiber nanocomposites utilizing 2D nanosheets with high electrical conductivity are poised to play a pivotal role in the prospective applications of wearable electronics and textiles.

- Flexible electronics

By integrating enhanced mechanical and electrical properties, both pristine and hybrid fiber nanocomposites based on 2D nanosheets find diverse applications across various domains. Cutting-edge uses primarily center around the realm of flexible and wearable electronics, leveraging the tensile strength, toughness, electrical conductivity, and flexibility inherent in fiber nanocomposites.

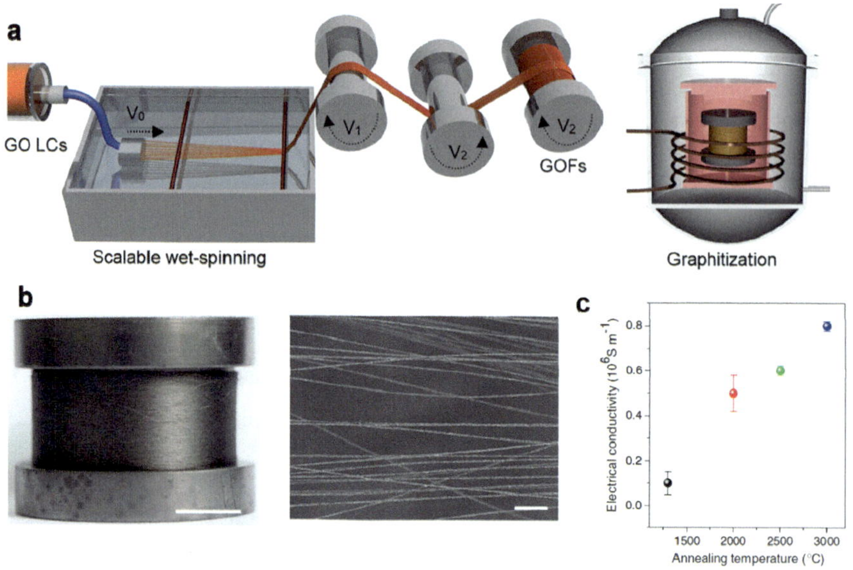

Fig. 18 **a** llustration of the preparation of graphene fibers via wet spinning followed by graphitization. **b** The reel with graphene fiber and SEM image. **c** Electrical conductivity of the graphene fiber [45]

– Wearable energy storage

Supercapacitors exhibit significant potential as energy storage devices for applications in compact, flexible electronics, boasting notable characteristics such as elevated capacitance, rapid charge–discharge rates, and dependable cycling stability. The swift advancements in science and technology necessitate supercapacitors to embody qualities such as lightweight design, portability, and ease of wear for optimal performance. In order to achieve their performance, the resulting fibers can be directly used as the electrodes for wearable supercapacitors. For instance, the graphene fibers were prepared through the integration of aligned CNTs with interleaved N-doped reduced GO nanosheets via wet spinning. This designed interconnected structure showcasing a substantial specific surface area (396 $m^2\ g^{-1}$) and an impressive electrical conductivity (102 $S\ cm^{-1}$). When incorporated into a micro-supercapacitor based on graphene fibers, the flexible supercapacitor demonstrated noteworthy specific volumetric capacities, reaching 300 $F\ cm^{-3}$ in a PVA/H_3PO_4 gel electrolyte (Fig. 19) [46]. Furthermore, MXene/CNT fiber, prepared by biscrolling with an impressive MXene loading of approximately 98 wt%, achieved exceptional specific volumetric of ~1100 $F\ cm^{-3}$, gravimetric of 428 $F\ g^{-1}$, and linear capacitance values of 118 $mF\ cm^{-1}$ in a 3 M H_2SO_4 electrolyte [31]

Moreover, fibers batteries represent a significant application domain for fibers nanocomposites based on 2D nanosheets in the realm of wearable energy storage.

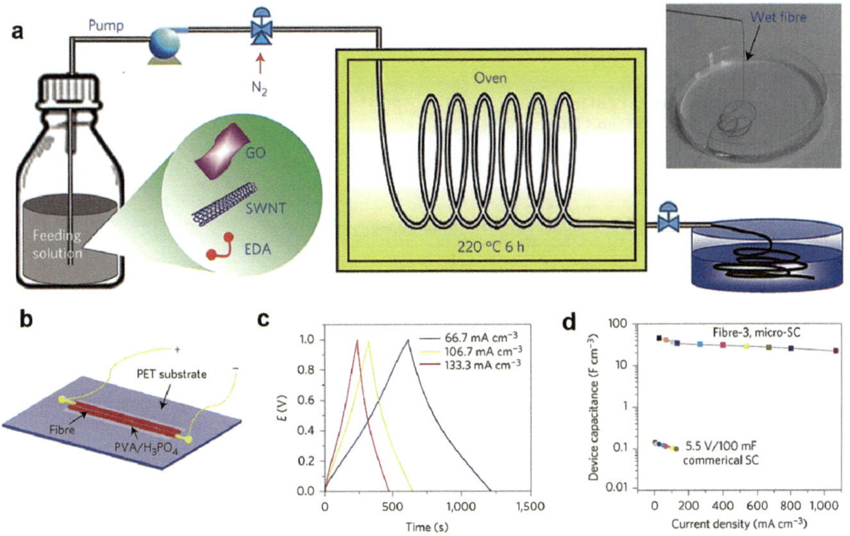

Fig. 19 **a** Illustration of the fabrication of graphene fibers via wet spinning. **b** Schematic of a micro-supercapacitors constructed using two graphene fibers. **c** Galvanostatic charge/discharge curves at various current densities. **d** Volumetric device capacitance at different current densities [46]

Such these fibers could be the good candidates for the lithium–sulfur battery, lithium-ion batteries, and so on [47]. Employing the wet-spinning technique, a graphene/CNT/S fiber was created to serve as the positive electrode with the negative electrode consisting of a lithium fiber (Fig. 20) [48]. Upon inserting a separator in between, the two types of fibers were amalgamated and enclosed within a plastic tube to form a fibrous lithium-sulfur battery. This battery exhibited a high capacity of ~1300 mAh g^{-1}, even with a specific capacity (2.49 mAh cm^{-2}). Notably, its discharge performance remained stable even after undergoing 30 cycles under bending tests.

- Wearable sensors

Thanks to the excellent electrical conductivity and exceptional flexibility inherent in fiber nanocomposites derived from 2D nanosheets, they are well-suited for application in wearable sensors, including wearable strain sensors, pressure sensors, and gas sensors [21]. For instance, a gel-like GO fiber around a glass rod, resulting in a helical texture that transformed into a helical graphene fiber spring through air drying and chemical reduction. The fiber spring exhibited an elongation of 480% and demonstrated robust mechanical durability. To enhance functionality, magnetic

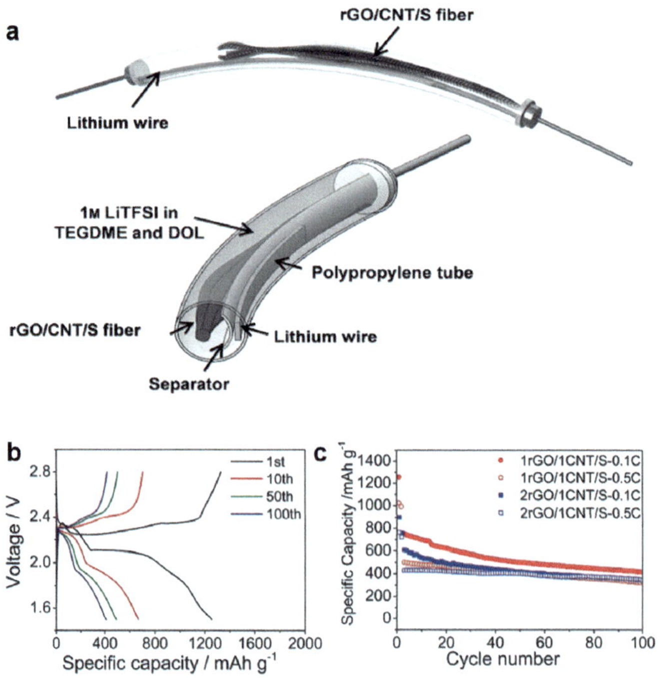

Fig. 20 **a** Depiction of the lithium-sulfur battery assembly using graphene fiber nanocomposites and a cross-sectional view of battery components. **b** Charge/discharge profiles recorded at 0.1 C. **c** Galvanostatic cyclic stability evaluation of 1rGO/1CNT/S and 2rGO/1CNT/S fiber electrodes over 100 cycles at 0.1 and 0.5 C [48]

ferric oxide was introduced into the fiber spring, enabling magnetic control of its stretching and compression. This magnetically controlled strain sensor finds practical application in magnetically controlled circuits [49]. A cotton textile with MXene coating was produced using the dip-coating and drying method (Fig. 21) [50]. Capitalizing on the textile substrate's undulating network structure and MXene's exceptional electrical properties, the resulting sensor exhibited notable sensitivity, registering ~12,000 kPa^{-1} for pressures at the range of 29 to 40 kPa. It showcased a rapid response time of 26 ms and remarkable stability across 5600 cycles. Furthermore, the MXene-textile sensor can achieve real-time monitoring of human activities, including wrist pulse, pronunciation, and finger movements of touching and bending. These tests indicate the potential of the textile sensor base on 2D nanosheets for integration into wearable and flexible electrics.

– Wearable energy harvesters

Energy harvesters play a vital role as wearable devices for converting energy, extracting electrical power from sources like mechanical strain energy (such as walking, bending, and running), thermal energy, sound, and light. Typically, these devices undergo conversion by responding to specific stimuli [21, 25].

For example, engineering helical microstructures by two GO gel fibers, the twisted graphene oxide fibers were formed to accumulate substantial strain energy in the process [51]. When the twisted fibers were exposed to moisture, it swiftly released this stored energy through interactions with hydrophilic oxygen functional groups, resulting in a rapid revolution at 5190 rpm. This actuation was robust, yielding

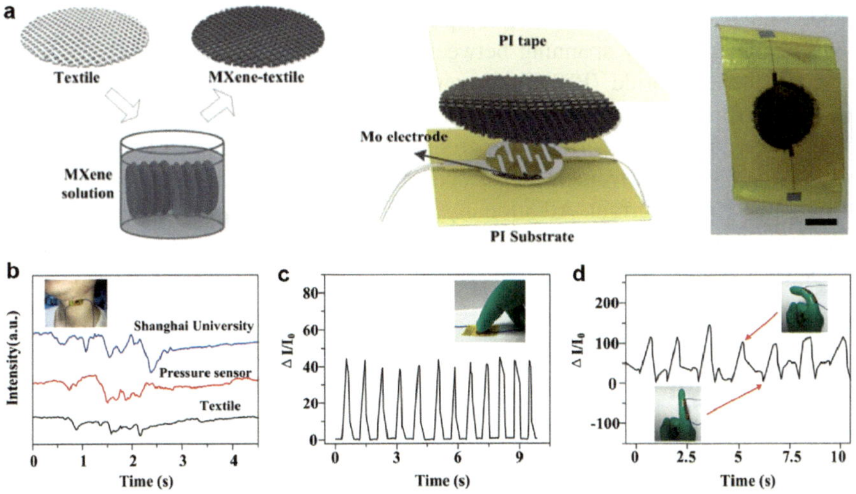

Fig. 21 **a** Illustration of the fabrication of MXene-textile and its pressure sensor via coating. **b** Current variations observed during the wearer's articulation of 'Shanghai University,' 'Pressure sensor,' and 'Textile.' Responsive curves resulting from finger touch **c** and finger flexion (**d**), respectively [50]

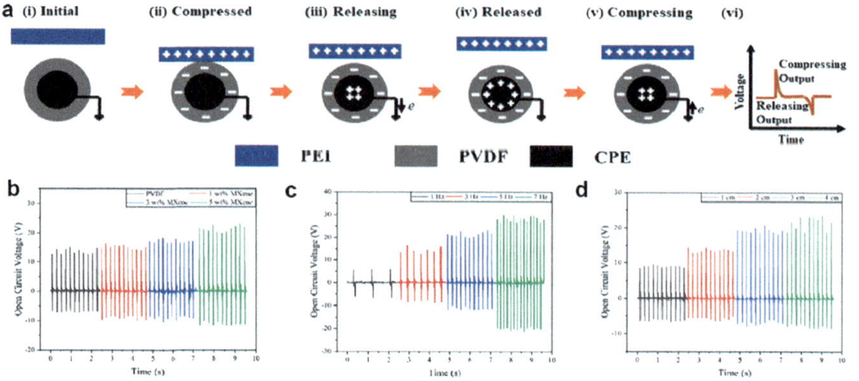

Fig. 22 **a** Single MXene electrode fiber working principle in triboelectric nanogenerators with a representative output for one contact-separation cycle. **b** Open circuit voltage comparison between pristine fibers and those embedded with MXene. **c** Open circuit voltage waveforms of nanocomposite fibers with 5 wt% MXene at varying frequencies (contact length: 3 cm). **d** Open circuit voltage waveforms (VOC) of nanocomposite fibers with 5 wt% MXene at different contact lengths (frequency: 5 Hz) [38]

a peak power output of 71.9 W kg^{-1}. Upon connecting fibers to the electromagnetic induction device, an open-circuit voltage (1 mV) and a short-circuit current (40 μA) were achieved. Meanwhile, some polar solvents like ethanol, methanol, and acetone induced fast responses [52]. Specifically, acetone-induced flexible fiber device produced a significant start-up torque (2.7×10^{-7} N m) and a record rotor kinetic power (89.3 W kg^{-1}). Moreover, through configuring the arrangement of handed fibers, the electrical power output varied from 9.8 to 3 W kg-1, offering a controllable frequency spanning between 10 and 5 Hz. In addition, the MXene/polyvinylidene fluoride (PVDF) fiber were scalably fabricated through thermal drawing to form triboelectric nanogenerators (Fig. 22) [38]. The synergistic interaction between the surface termination groups of MXene and the polar PVDF polymer results in a 53% improvement in the open-circuit voltage and a 58% improvement in the short-circuit current. Additionally, the fabric exhibits a power density of 40.8 mW m^{-2} at the matching load of 8 MΩ, maintaining stable performance throughout 1.2×10^4 cycles.

3 Conclusion

This chapter discuss that the incorporation of 2D nanosheets into fibers enhances their mechanical strength, toughness, electrical conductivity, and flexibility via various preparation approaches, thereby opening up opportunities for innovation in various fields. The fiber nanocomposites have demonstrated the successful development of these nanocomposites, showcasing their potential in areas such as wearable electronics, flexible devices, and wearable energy storage systems. The unique combination of improved mechanical performance and electrical abilities in these nanocomposites has facilitated their integration into flexible and wearable electronics.

Looking ahead, the continued exploration of fiber nanocomposites based on 2D nanosheets holds great promise for advancements in materials science and engineering. Researchers are expected to focus on tailoring the properties of these nanocomposites for specific applications, optimizing their performance, and exploring novel synthesis methods via controlling the interface and structures. The development of scalable and cost-effective manufacturing processes will be crucial for the widespread adoption of these materials. Moreover, efforts to enhance the environmental sustainability of the production processes and the recyclability of the resulting nanocomposites will likely be a key area of future research. The integration of smart functionalities, such as human–computer interaction capabilities for internet of things, into these fiber nanocomposites is another exciting avenue for exploration. This could lead to the creation of intelligent textiles and wearable devices with enhanced performance and functionality.

In conclusion, fiber nanocomposites based on two-dimensional nanosheets have already demonstrated their potential in diverse applications. As research in this field continues, we anticipate witnessing further breakthroughs, pushing the boundaries of what is achievable with these innovative materials. The synergy between materials science, nanotechnology, and engineering is poised to unlock new possibilities, making these nanocomposites integral components in the next generation of advanced materials.

References

1. Reeder JT, et al. Science. 2022;377:109.
2. Lee C, et al. Science. 2008;321:385.
3. Naguib M, et al. Adv Mater. 2011;23:4248.
4. Manzeli S, et al. Nat Rev Mater. 2017;2:17033.
5. Youngblood N, et al. Nat Photonics. 2015;9:247.
6. Nakata S, et al. Nat Electron. 2018;1:596.
7. Ates HC, et al. Nat Rev Mater. 2022;7:887.
8. Ma C et al. Adv Funct Mater. 2021;2009524.
9. Levitt A et al. Adv Funct Mater. 2020;2000739.
10. Lee SH, et al. ACS Appl Mater Interfaces. 2020;12:10434.
11. Park JW, et al. MRS Commun. 2019;9:114.
12. Chen S, et al. Nano Energy. 2015;15:642.
13. Guo K, et al. Small. 2016;12:1024.
14. Xu Z, Gao C. Nat Commun. 2011;2:571.
15. Park H et al. Sci Adv. 2018;4(eaau2104).
16. Cruz-Silva R, et al. ACS Nano. 2014;8:5959.
17. Yaman M, et al. Nat Mater. 2011;10:494.
18. Schmidt MA, Argyros A, Sorin F. Adv Optical Mater. 2016;4:13.
19. Zhou T, et al. Nat Commun. 2022;13:4564.

20. Hou J, et al. Opt Exp. 2008;16:5983.
21. Abouraddy AF, et al. Nat Mater. 2007;6:336.
22. Tao GM, et al. Int J Appl Glass Sci. 2012;3:349.
23. Sorin F, et al. Adv Mater. 2007;19:3872.
24. Gumennik A, et al. Adv Mater. 2012;24:6005.
25. Fang B et al. Adv Mater. 2019;1902664.
26. Xin G, et al. Science. 2015;349:1083.
27. Naficy S, et al. Mater Horiz. 2014;1:326.
28. Shin H, et al. ACS Nano. 2021;15:3320.
29. Jiang Q, et al. Adv Energy Mater. 2018;8:1703043.
30. Levitt AS, et al. J Mater Chem A. 2019;7:269.
31. Wang Z, et al. Small. 2018;14:1802225.
32. Shahzad F, et al. Science. 2016;353:1137.
33. Pei S, et al. Nat Commun. 2018;9:145.
34. Gu J, et al. Adv Mater. 2023;35:2209527.
35. Li S, et al. ACS Nano. 2021;15:7821.
36. Zhang Y, et al. ACS Nano. 2018;12:8901.
37. Wang X, et al. ACS Nano. 2018;12:12638.
38. Hasan MM. Small. 2022;2206107.
39. Zhang Y, et al. Adv Mater. 2016;28:2834.
40. Zhou T et al. Adv Mater. 2023;2305807.
41. Lipatov A et al. Sci Adv. 2018;4(eaat0491).
42. Uzun S, et al. Adv Funct Mater. 2019;29:1905015.
43. Zhnag J, et al. ACS Cent Sci. 2020;6:254.
44. Xin G, et al. Nat Nanotech. 2019;14:168.
45. Xu Z, et al. Adv Mater. 2016;28:6449.
46. Yu D, et al. Nat Nanotech. 2014;9:555.
47. Zhang Y et al. Angew Chem Int Ed. 2014;126, 14792.
48. Chong WG, et al. Adv Funct Mater. 2017;27:1604815.
49. Cheng H, et al. Nanoscale. 2014;6:11052.
50. Li T, et al. J Mater Chem C. 2019;7:1022.
51. Cheng H, et al. Adv Mater. 2014;26:2909.
52. Fang B, et al. Mater Horiz. 2019;6:1207.

Chapter 9
Fiber-Based Wearable Sensors with Machine Learning

Ziwang Luo, Jiaming Xu, Xuhui Zhou, Haozhe Zhang, Zhixun Wang, Zhiping Lin, and Lei Wei

Abstract When people discuss science and technology, they instinctively inquire about the practical applications. This book primarily focuses on optical fiber and optoelectronic fiber. Previous chapters have already covered some direct applications of these technologies. In this chapter, we will delve into optical fiber sensors and a specialized area involving wearable fiber sensors that utilize machine learning. The association between optical fiber and wearable sensors primarily stems from the fibrous nature of optical fiber. Fiber is the raw material of people's daily clothes. Fiber sensors with fiber as the base are naturally the first choice for wearable sensors. As the most powerful data processing method in recent years, machine learning has quickly gained popularity in the field of wearable devices. As a result, fiber sensors, wearable devices, and machine learning are connected together. To provide a more comprehensive perspective, this chapter will not only examine wearable sensors based on optical fiber but also explore electrical-based fiber sensors and summarize the current applications of machine learning, a rapidly growing field, in fiber sensors.

1 Introduction

Intelligent wearable devices are increasingly important in our lives due to their convenience, intelligence, and inconspicuousness. They have found extensive applications in healthcare, sports, and environmental detection. Current wearable systems consist of three essential parts: a sensing device, an information transmission channel, and a data processing unit. In the case of wearable sensors, regardless of the changing target information, they have to be worn by human beings; thus, comfortability is of great importance. However, traditional electronic sensors are lack of the desired

Z. Luo · J. Xu · X. Zhou · H. Zhang · Z. Wang · Z. Lin (✉) · L. Wei (✉)
School of Electrical and Electronic Engineering, Nanyang Technological University, Singapore, Singapore
e-mail: ezplin@ntu.edu.sg

L. Wei
e-mail: wei.lei@ntu.edu.sg

L. Wei (ed.), *Advanced Optical and Optoelectronic Fibers*, Advances in Optics and Optoelectronics, https://doi.org/10.1007/978-981-97-6218-7_9

level of comfortability due to the inherent rigidity of silicon wafers and metallic materials, thus users can easily feel the presence of wearable devices. Consequently, such conventional sensors cannot be integrated seamlessly into everyday clothes. In the context of daily wearables such as hats, masks, shirts, skirts, trousers, socks, shoes, etc., the predominant material choice is fabric. Fabrics can perfectly fit people's pursuit of comfortability and the extensibility of wearable items. Therefore, developing new fiber sensors and constructing wearables using fiber-based sensors have emerged as attractive and feasible topics within the field of wearable devices.

Apart from fiber sensors, wearable devices face another challenge in terms of data processing. Before the widespread use of machine learning, the main data processing methods were threshold-based analysis, statistical analysis, and signal processing techniques. The threshold-based analysis compares the data collected by sensors against pre-set thresholds. For instance, a heart rate monitor might flag any reading above or below certain beats per minute as abnormal. Statistical measures include means, medians, variances, and standard deviations. For example, calculating the variability in heart rate. The signal processing techniques cover a wide range of methods, including filtering, time series analysis, Fourier transform, and wavelet transform. They all aim to avoid interference and extract essential patterns. Although highly versatile, threshold-based and simple statistical analyses lack the control over details. Signal processing enhances the discovery of patterns, but it has major limitations. For example, different data patterns vary, which usually require manual search, which is time-consuming and laborious. In complex signals, specific patterns may even overlap with other information and be covered by trivial signals, further increasing the difficulty of extracting important information.

Machine learning joins the queue of data processing methods thanks to the rapid advancement of artificial intelligence within the last ten years. Machine learning and its important branch, deep learning, first of all, eliminates the need for the manual search for patterns or features, which greatly improves efficiency. Secondly, deep learning's insight into details is beyond the reach of all the above methods. Thirdly, massive data can be a burden for conventional methods, but they suit machine learning well. Machine learning's ability to uncover valuable insights and patterns within the data makes it an excellent tool for dealing with complex and sophisticated human body activities among massive data generated by wearables. All in all, data processing and analysis are the steps after sensor data collection and before the wearable device takes action. This step determines the intelligence of the wearable device. In the past, most people used statistical and signal processing techniques at this step, but now machine learning is gradually replacing statistical and signal processing techniques because of its excellent and effective information extraction and analysis capabilities, becoming a must for the new generation of wearable devices to analyze information.

This chapter focuses on intelligent wearable devices with the successful fusion of machine learning and fiber-based sensors. We start with a summary of optical fiber sensors. Optical fiber sensors are categorized using two methods: fiber dimension and working principle (Fig. 1). According to the dimension of the fiber, sensors are divided into macroscopical and microscopical fibers. Macroscopical fibers are of size in centimeters appearing as fibers in the usual sense, while the characteristic dimensions of microscopical fibers are usually in microns or nanometers. Further subcategories within macroscopical fibers include single-fiber and multiple-fiber structures. On the other hand, microscopical fibers are classified into surface coverage and percolation. Surface coverage means that fibers are sprayed onto the surface of a base material, and percolation means the fibers are merged into base materials. When the working principle is considered, sensors will be categorized into optical sensors and electronic sensors. Optical sensors encompass four groups: fiber Bragg grating (FBG), interference image, intensity variation, and Fabry-Pérot interferometer, while electronic fiber-based sensors, as a supplement to this chapter, include three groups: piezoresistive, triboelectric, and chip-in-fiber. After the summary of wearable fiber sensors, an introductory overview of the machine learning algorithms will be presented. It is divided into two subsections: traditional machine learning algorithms and deep learning algorithms. The traditional machine learning algorithms cover linear regression (LR), k nearest neighbors (KNN), support vector machine (SVM), random forest, XGBoost, and K means clustering. Meanwhile, the deep learning algorithms include multilayer perceptron (MLP), convolutional neural network (CNN) and ResNet, recurrent neural network (RNN) and long short term memory (LSTM), and transfer learning. Finally, challenges and potential directions for future research are provided in the Conclusion.

2 Optical Fiber-Based Sensors

Optical fiber sensors use light as the medium to respond to target information. It is divided into four categories according to their working principles within the scope of this chapter: Fiber Bragg Grating (FBG), interference image, intensity variation, and Fabry–Perot interferometer.

- 1 Fiber Bragg Gratings (FBGs)

An FBG is a segment of optical fiber with a periodic variation in the refractive index of the fiber core [1]. According to the Fresnel reflection theory, light traveling between media of different refractive indices may reflect and refract at the interface. Therefore, such a periodic variation in the refractive index reflects light in phase with the grating period and transmits all others. The first fiber Bragg grating was demonstrated by Ken Hill in 1978. The wavelength of the light reflected is determined by the following equation $\lambda_B = 2n_e\Lambda$, where λ_B is the wavelength of the light reflected, n_e is the effective refractive index, and Λ is the grating period. Interestingly, when the optical fiber is deformed, the grating period changes, resulting in a shift in the wavelength of the reflected light, as illustrated in Fig. 2a. Fiber sensors based on FBG utilize this property to detect deformations in the target area.

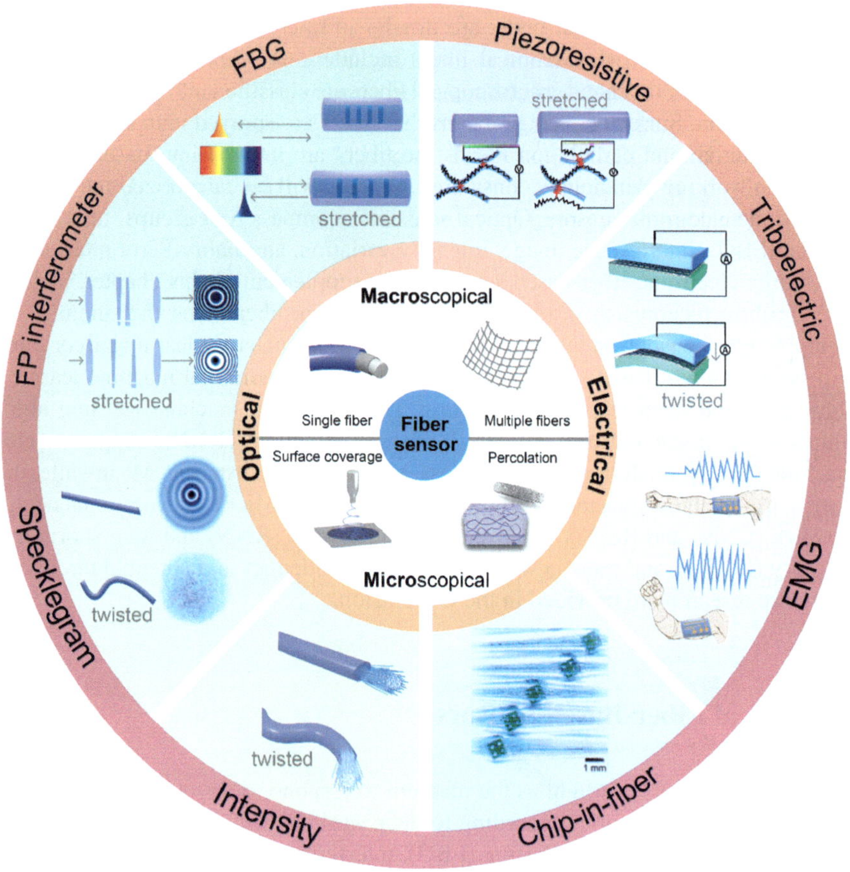

Fig. 1 Categorization of wearable fiber sensors. The inner circle classifies fiber sensors into 'Macroscopical' and 'Microscopical' according to the fiber dimension. The outer pie chart shows the classification according to the working principles

– 1.1 Single Fiber FBGs

As mentioned above, FBG can detect the deformation of the fiber. However, how many FBGs and how many fibers containing FBGs to use is a problem that designers need to consider. The most straightforward solution is to directly use a single fiber containing a single FBG. For example, single FBG was used to sense helmet blunt force impact [2]. There was only one optical fiber with one FBG inscribed (Fig. 2b). When the blunt force came from different directions with different strengths, the helmet deformed in different patterns so that the wavelength shifts of reflected light also fluctuated distinctly (Fig. 2c). These wavelength shifts versus time graphs were

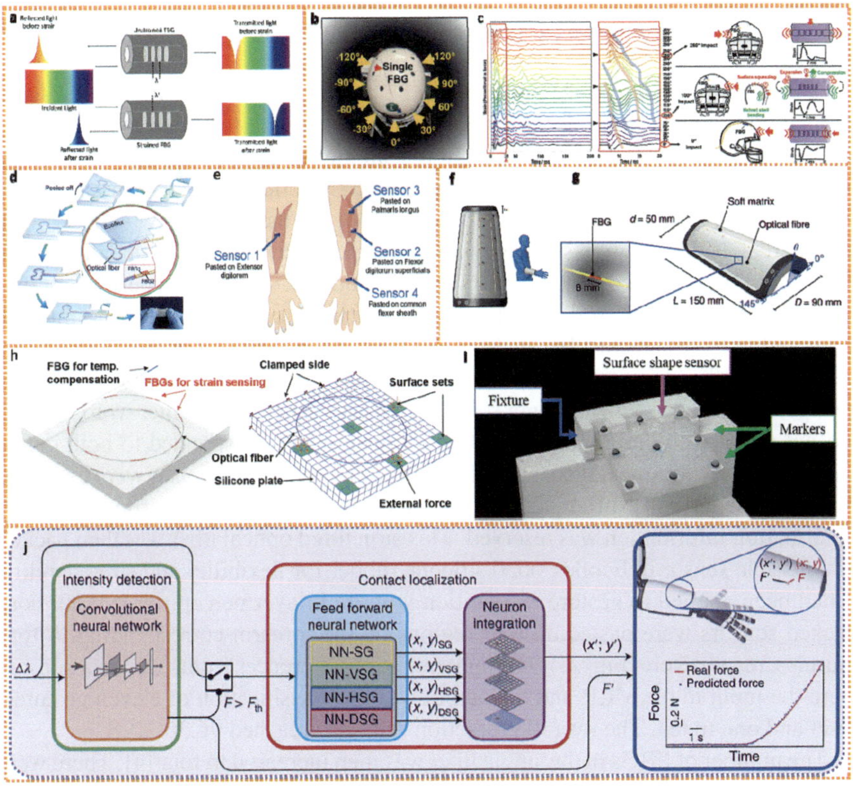

Fig. 2 **a** Working principle of the FBG. **b** The placement of the FBG inside the helmet and some possible hit angles. **c** The wavelength variation spectrum: each curve describes the variation of wavelength with time at a specific hitting angle. **d** The structure and the preparation process of the strain sensor in the shape of Lindernia nummulariifolia. **e** The arrangement of sensors on the forearm for gesture recognition. Reproduced with permission. [3] Copyright 2022, American Chemical Society. **f** The arrangement of 16 FBGs in the sensitive skin patch on the forearm. **g** The characteristic dimensions of the sensitive skin patch. **h** Left: an overview of the surface shape sensor structure: a single fiber is looped in two circles away from the midplane, and each circle contains 8 FBGs. Right: Finite element model of the sensor. Twelve surface sets (6 shown in green) are predefined to apply random pressure loadings to simulate external forces. **i** Experimental setup for training data collection. The sensor is covered with 9 IR-tracking markers and clamped on a fixture. Reproduced with permission. [6] Copyright 2019, IEEE. **j** Deep learning neural network structure and results: A CNN is developed for force intensity detection. For forces above the 50 mN threshold, a feedforward neural network and subsequent multigrid neuron integration process (NIP) implement the localization of the contact of the load applied onto the skin surface

then input to SVM and KNN to determine the magnitudes, directions, and latitudes of the blunt force applied to the helmet by analyzing signals from just one FBG. For each index, the model that presented the best performance was selected. SVM achieved a mean absolute percentage error (MAPE) of 8.991% for magnitude prediction. KNN achieved a MAPE of 4.67% for direction determination and 6.373% for latitude prediction.

Although a single FBG can determine the magnitude and direction of the force applied on the fiber, multiple FBGs are commonly needed to obtain more comprehensive information or collect information from multiple points simultaneously. We need to notice that when multiple FBGs are placed in series, each FBG should reflect the light of a specific wavelength different from all the others. Two FBGs were used to realize deformation sensing without interference from environmental factors (Fig. 2d) [3]. It proposed to bend the optical fiber into the shape of Lindernia nummulariifolia so that larger strain and bending angle measurements could be achieved. Both ends of the optical fiber had an FBG inscribed. FBG1 was configured as a reference, so the intensity of its reflection spectrum only depended on environmental changes like temperature and light source intensity variations. In other words, FBG1 did not respond to deformations. FBG2, on the contrary, responded to both deformations and environmental changes. Therefore, environmental noises were excluded by taking the intensity difference of FBG1 and FBG2 reflection spectrums, and only deformation information was reserved. This structured optical fiber was then packed as a flexible sensor in Ecoflex 0030 silicone rubber for flexibility and stretchability. It had been applied to gesture recognition with multilayer perceptron (MLP). Four packed sensors were pasted on four regions on the forearm corresponding to four muscles, respectively (Fig. 2e). The light intensity differences from these four sensors were the input to the MLP, and the output was the classification of eleven gestures: reset and one to ten. The overall prediction accuracy reached 96.36%.

The number of FBGs in the single fiber was then increased to four [4]. There were 4 FBGs in the optical fiber, embedded straightly in the flexible substrate. When a force was applied on the surface, these four FBGs were stretched, leading to shifts in the wavelength of reflected lights. These 4 wavelength shifts were collected and analyzed by an MLP for indentation location and force strength determination. For the 1 mm thick sensor, the RMS error of displacement prediction was 10.73 mm, and the force RMS error was 0.69 N. For the 6 mm thick sensor, the RMS error of displacement prediction was 2.45 mm, and the force RMS error was 0.164 N.

Similarly, a single fiber with multiple FBGs was also used to determine indentation location and force strength [5]. However, the structure constructed was more complex: 16 FBGs were inscribed in series to mimic Ruffini receptors of human skin. Mechanoreceptors of human skin can be classified into four types: surface-located fast adapting, surface-located slow adapting, deeply located fast adapting, and deeply located slow adapting. Surface-located receptors can capture smaller and more well-defined contacts than deeply located receptors. Ruffini receptor is the deeply located slow-adapting mechanoreceptor in human skin. 16 FBGs were inscribed in one optical fiber which was then embedded in an artificial skin frame to realize touch sensing. Every FBG was designed to reflect a specific wavelength. The arrangement of the optical fiber was shown in Fig. 2f, and the characteristic dimensions were shown in Fig. 2g. Touching one spot induced wavelength shifts of multiple neighboring FBGs. Sixteen wavelength shifts were fed into a CNN model for force intensity detection, followed by four feedforward neural networks and a neuron integration for contact localization. The feedforward neural network was an MLP without backpropagation, and neuron integration calculated the weighted mean of four feedforward neural networks. Figure 2j illustrates the structure of this machine-learning algorithm. The test set gave a median error of 35 mN for force prediction and 3.2 mm for position prediction.

An interesting structure was assembled to reconstruct the surface shape [6]. The fiber that contained 16 in-series FBGs was looped in two circles, one on top and the other at the bottom, embedded in an elastomeric silicone plate, as shown in Fig. 2h. One side of the plate was clamped, and nine grid points were chosen (four at the corners, four at the middle of borders, and one at the center) (Fig. 2i). While the plate was pressed, 16 FBG wavelength shifts were collected to predict the displacements of the chosen nine grid points by using an MLP. This design achieved a root mean square (RMS) error of 1.17 mm for reconstructed node displacements.

– 1.2 Multiple Fiber Interweaves

Although a single fiber can realize a simultaneous collection of information by inscribing multiple FBGs in series, all the FBGs in the same fiber are strongly correlated. However, information needs to be collected independently. For example, when a cylindrical object is bent, the outer contour is stretched more than the inner contour. If both tensile lengths need to be recorded independently, embedding two parallel optical fibers is more feasible than one twisted fiber. Therefore, in order to collect information from various sections separately and simultaneously, multiple fibers interweave is necessary. A respiratory flow prediction wearable system was designed using two parallel optical fibers (two FBGs in each fiber) and an M-IMU (including an accelerometer and a gyroscope) [7]. The M-IMU was placed around the upper thorax. The first optical fiber containing FBG1 and FBG2 was fixed around the lower thorax, while FBG3 and FBG4 embedded in the second fiber were set around the abdomen, as illustrated in Fig. 3a, b. Both optical fibers were elongated during inhalation, resulting in wavelength shifts of reflected light. In the end, four wavelength shifts, acceleration, and angular velocity were fed into a transfer learning LSTM model to predict respiratory volume, as shown in Fig. 3k. The best root mean square error was 9.97 L/min.

Similarly, three parallel optical fibers were encapsulated together, each fiber had three FBGs inscribed in series, in a Polydimethylsiloxane (PDMS) compliant polymer as a force strength estimator (Fig. 3c, d) [8]. The strain experienced by each FBG was reflected by the Bragg wavelength shift. The wavelength shifts from 9 FBGS were the input to an MLP for the computation of the sensor's inverse characteristics. A 21 mN root mean square error (RMSE) was achieved in determining the force over the 8 N experimented load range.

The same proposal was made to sense the shape of continuum robots: multiple parallel FBG fibers along the contour of the tool body [9, 10]. The green circle in Fig. 3e shows the cross-section of this FBG shape sensor: four fibers in total, one at the center, and the rest three were distributed evenly around the center [9]. Traditional shape reconstruction approaches require tedious characterizations of many parameters and must know the precise location relative to the centerline. The FBG shape

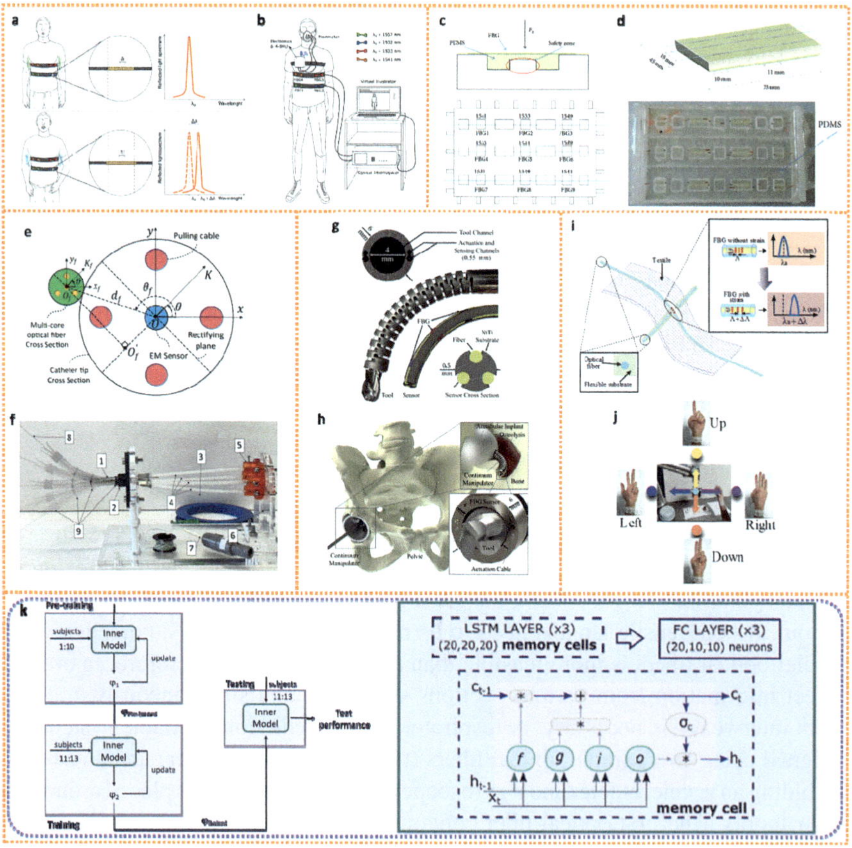

Fig. 3 **a** Schematic diagram of FBG placement and working principle. **b** An overview of the experiment setup. **c** Cross section of a strain sensing unit. **d** The strength estimator's characteristic dimensions and the actual appearance. **e** Cross-sectional view of the catheter tip and the multicore FBG fiber. **f** Fluidics-driven catheter system for data acquisition and algorithm validation. Reproduced with permission [9] Copyright 2023, IEEE. **g** Continuum manipulator tip cross-sectional view showing the actuation, sensing, and tool channels. **h** Robot-assisted treatment of pelvic osteolysis. Reproduced with permission. [10] Copyright 2021, IEEE. **i** Schematic diagram of flexible optical crossbar sensor. **j** Movement direction of the robot arm corresponding to different gestures. Reproduced with permission. [11] Copyright 2022, IEEE. **k** Transfer learning algorithm and data processing workflow

sensor overcomes these two limitations and can be placed off-center. A two-degree-of-freedom benchtop fluidics-driven catheter system verified this idea (Fig. 3f). In the experiment, 20 wavelength shifts from this 4-core fiber during robot bending were sent to a three-hidden-layer MLP. This MLP achieved an average shape-sensing error of 0.24 and 0.49 mm in free-space and constrained-space experiments, respectively.

Similarly, a surgical continuum manipulator was created using three optical fibers (Fig. 3g, h) [10]. Twelve wavelength shifts from three optical fibers were the input to either a linear regression model or an MLP model. For MLP, two sets of input were used, including wavelength shifts at just one timestamp and wavelength shifts at several consecutive timestamps. Among all experiments, the MLP with input at a single timestamp outperformed the rest models, giving a maximum shape deviation error of 1.95 mm, where the shape deviation error was defined as the mean of absolute difference of all the points along the manipulator compared with ground truth.

Because of a parallel structure, there was naturally a vertical structure. However, the function of a vertical structure was not simultaneous sensing at different regions anymore. It was designed to emphasize the deformation at the intersection. Two fibers were vertically overlapped, and each had an FBG inscribed in the middle of the fiber so that one FBG was right on top of the other (Fig. 3i) [11]. This crossbar structure was weaved into a waistband and used for hand gesture recognition and further control of a robot arm (Fig. 3j). While gesturing, waist muscles contracted, and bones displaced slightly. This subtle information can be conveyed by the two wavelength shifts and decoded by an MLP, achieving an accuracy of 97.02% for a four-gesture recognition task.

- 2 Specklegram or Interference Image

In a multimode optical fiber, lights travel in different modes along various paths. They interfere with each other and show specific patterns [12]. Figure 4a shows an exemplary interference phenomenon along a multimode fiber. At the fiber end-facet, a 2D intensity image, called specklegram, can be captured by cameras (Figure 4b). When the multimode optical fiber deforms, interference changes due to the light path variation, and therefore the interference image or specklegram captured at the end-facet also alters. Considering the large number of eigenmodes in the fiber, rigorous conventional analysis, like the beam propagation method, becomes infeasible. Hence, machine learning is used to relate image variations to actual fiber deformations.

Since the specklegram is in the form of images, convolutional neural networks are normally used to extract the information. A system that is able to determine the location of perturbations along a multimode plastic optical fiber was designed (Figure 4c) [13]. A fiber hitter controlled by a linear stepper motor hit the fiber at different places as planned, and the CCD camera recorded the videos of speckle patterns. These videos were preprocessed and then used to train a CNN to figure out the perturbation location. The results showed a classification accuracy of 99% when there were three different locations, and this number decreased to 71% when the number of locations rose to ten.

Later, an improvement was made: except for the part used to perceive changes, the rest used single-mode fiber (SMF) to reduce noise [14, 15]. Scenarios were first simplified to direction changes where only a single point on the multimode fiber of the optical fiber was bent [14]. The light was incident into a single-mode fiber and then coupled to a multimode fiber (Figure 4d, e). The single-mode fiber side was clamped on a displacement stage, which could move in four directions: up, down, left, and right. The multimode fiber side was fixed on a CCD camera for image reception. In each direction, 10 points were selected in the step of 0.8mm. There were 40 points in total, corresponding to 40 classes for classification. Over 12800

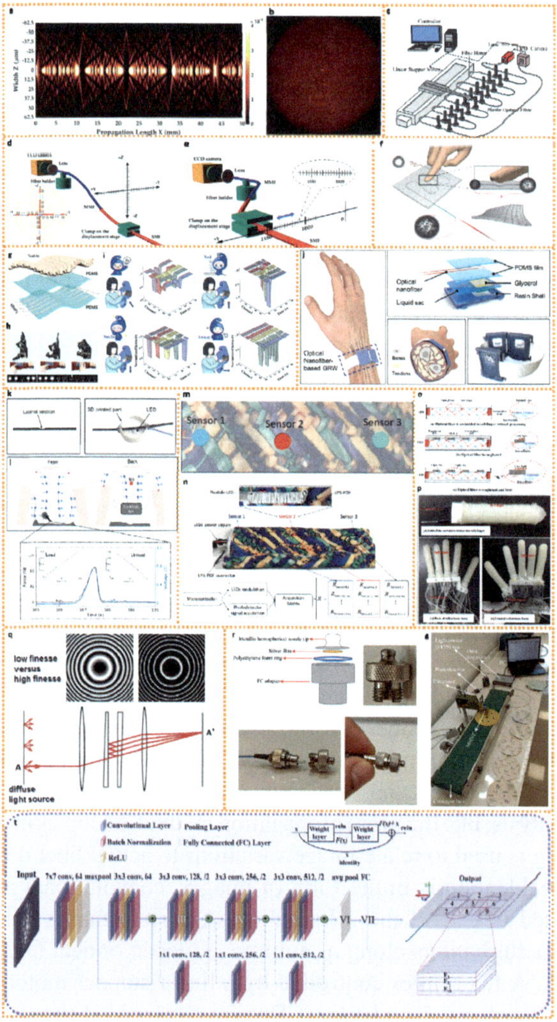
Propagation Length X (mm)
Width Z (μm)
Sensor 1
Sensor 2
Sensor 3
low finesse
versus
high finesse
diffuse
light source
Optical
nanofiber
Liquid sac
PDMS film
Glycerol
Resin Shell
Convolutional Layer
Pooling Layer
Batch Normalization
Fully Connected (FC) Layer
ReLU
Input
Output

◂**Fig. 4** **a** Intermodal interference profile inside the multimode fiber along the fiber axis. **b** Camera shot of light intensity distribution after fiber perturbation. **c** Set up used for automatically recording videos of the speckle pattern while perturbating a multimode plastic optical fiber at ten different locations. Reproduced with permission. [13] Copyright 2018, IEEE. **d** Stage movement pattern for directional sensing. Ten classes of positions are formed along each of the four axes with a span of 8 mm. **e** The fine spatial resolution within a small range along the + Y axis. **f** Soft tactile sensor with one fiber inside as transducer. Inset highlighting the effect of the applied load. Reproduced with permission. [15] Copyright 2020, Elsevier Ltd. **g** Structure of optical smart textiles (OST). **h** Logic control of a robotic hand. **i** Hand slide motions and corresponding light intensity output of five MNFs. Reproduced with permission. [16] Copyright 2022, Donghua University. **j** Overview of the gesture-recognition wristband (GRW): Configuration, Schematic illustration of a single nanofiber-based pressure-sensor unit, Cross-section of the wrist with GRW, Photograph of the GRW. **k** Removal of part of the fiber material creates a lateral section, and the LED is coupled to the fiber lateral section. **l** Photonic Smart Garment overview and the sensor response when a force is applied on the top of the sensor. **m** Force map from the sensors' responses with force applied on the textile. **n** A picture and schematic representation of the proposed multifunctional smart textile. Figure also shows the acquisition matrix with the responses of the sensors. **o** Measurement principle of optical fiber-based curvature sensor (OFCS). **p** Soft robotic hands with OFCSs. **q** Working principle of Fabry–Pérot interferometer. **r** Structure of roughness sensing tip. **s** Schematic diagram of the experimental setup. Reproduced with permission. [22] Copyright 2021, Elsevier B.V. **t** Structure of simplified ResNet with ten layers. Inset shows a building block in residual learning

interference images were fed into a CNN-based model called ResNet-18 for training. The input was the interference image, and the output was the location of the fiber tip, including the direction and displacement distance. The final test accuracy reached 97.34%.

The work mentioned above only investigated the deformations that happen at a single point on the optical fiber, which was oversimplified. Thus, the structure of the sensor and the sensor applied to tactile sensing were later improved (Figure 4f) [15]. The fundamental structure was still the same: the light was incident into a single-mode fiber at the beginning and then coupled to a multimode fiber. The output of the multimode fiber was connected to the camera. But this time, the multimode fiber part did not go straight forward and only captured directional information, instead, it winded like a snake forming an S-shape path and the whole section felt the force from above. This design was more practical and made more use of the ResNet. To make the sensing part flexible, a small section of SMF, the connection between SMF and MMF, and the winding part of MMF were all embedded in a flexible silicone pad. During data acquisition, the pad was divided into nine equal square sections, and they were pressed with different strengths individually. The machine learning algorithm used was still ResNet (Figure 4t). After training, given a new interference image, the model could tell the contact location and the corresponding force strength applied to the area. In different experiment setups (film thickness, number of training images), a positioning accuracy ranging from 0.8456 to 0.99 and a force accuracy ranging from 0.8042 to 1 were achieved.

- 3 Light Intensity Variation

Apart from FBG and interference images, light intensity is a more straightforward medium. Whenever the fiber deforms, the power transmission inside the fiber is inevitably influenced. Although the mechanism is straightforward, conveying complex information using light intensity variation is challenging. Some scholars decide to use optical micro/nanofibers (MNFs). Due to MNFs' small diameters, when undergoing significant deformation caused by forces, MNFs will produce significant changes in light transmission loss, causing obvious changes in light intensity [16, 17]. Some innovative changes are made to the optical fiber, so that deformation information can be effectively captured by investigating light intensity variations [18–20].

In one article, the researchers placed five MNFs in parallel, encapsulated them with PDMS, and then adsorbed them under the fabric (Figure 4g) [16]. Fabrics in this structure were put into two applications: logic control of a robotic hand (Figure 4h) and gesture recognition (Figure 4i). Logic control of a robotic hand was to press different numbers of optical fibers at different positions to control the robotic arm to make different actions. Gesture recognition was realized by sliding the human hand in different directions on the surface of fabrics, producing five channels of light-intensity information. The author used multi-class SVM to judge the hand's action. The light intensity of five fibers was used as input. The multi-class SVM made decisions among eight actions, achieving a recognition accuracy as high as 98.1%.

The second article using MNF was also about gesture recognition, but unlike human hands rubbing on the surface of the device, this article embedded fiber into the bracelet (Figure 4j) [17]. In each sensing unit, the optical nanofiber was bent into a U shape and embedded in PDMS, and a piece of glycerol was encapsulated with a liquid sac under the PDMS film to transfer the pressure to the optical nanofiber as losslessly as possible. According to the tendon distribution of the human wrist, the article assigned three sensing units to the bracelet. The light intensity data collected by these three sensing units were consolidated and then used as input to SVM for gesture recognition. The article recognized 12 gestures with an accuracy rate of 94%.

In addition to using MNF, modifications to ordinary optical fibers can also achieve the goal of collecting information more effectively. A photonic smart garment was designed with 30 intensity variation sensors distributed in 4 polymer optical fibers (Figure 4k, l) [18]. Different from common endpoint light sources, sensors in this work had one LED for each of them. At the position of each sensor, half of the protective jacket was removed, and the exposed core was aligned with the LED light source. Each core-LED pair was encapsulated by clear urethane rubber (Figure 4k). In this case, any stress or strain applied on the assembled optical fiber would induce optical power variations at two ends of the fiber. To avoid intersensory noise, LEDs flashed sequentially, and there was only one bright LED at any time. The receiver at the end of the fiber was synchronized, so there was a match between time stamps and sensor labels. The transmitted power spectrum was then sent to a KNN classifier for movement analysis, including walking, squatting, running, etc. The overall accuracy reached 94%. This smart garment could also monitor the cadence which is the rate of movement. The walking rate, running rate, and breathing rate were all tested with the error rates below 2.22%.

This strategy was also applied in another work: along the so-called LPS-POF optical fiber, three LEDs were side-coupled to the fiber as the light source, and the end faces of the fiber were connected to photodetectors (Figure 4m, n) [19]. Then, an on-off keying (OOK) modulation scheme was applied to three LEDs, which means that LEDs flashed one by one, and only one LED was activated at a time. Both LEDs and optical fiber were embedded in neoprene textile fabric, forming a stretchable textile for multimodal wearable sensing. As temperature, transverse force, and angular displacement at different planes changed, the received light power varied. This textile was then positioned on the lower back of the volunteer to realize activity detection. The power variations were first processed by principal component analysis (PCA) to extract features and then got classified into three classes: walking, squatting, and sitting by K means clustering. The accuracy seemd to be close to 100%, extracted from the provided figure.

In addition to the above method, making the fiber more lossy by roughening the fiber with a laser also works (Figure 4o) [20]. When the fiber was straight, a small part of light escapes; while the fiber was bent, around half of the light was lost. This roughened fiber was embedded into the artificial finger along the finger contour (Figure 4p). Therefore, the light intensity transmitted through the fiber varied tremendously during gesturing. Apart from optical sensors, pressure sensors were also embedded. By feeding light-intensity spectrums and pressure information to machine learning algorithms, gestures or object properties like shape, size, and weight could be recognized. KNN, SVM, KMC, and LR were all tested. They all achieved at least 85% accuracy, and KNN performed the best: 97.96% for gesture recognition, 98.06% for object shape, 95.17% for object size, and 100% for weight.

- 3.4 Fabry–Perot Interferometer

Fabry-Perot (FP) interferometer is an optical cavity made from two parallel reflecting mirrors [21]. The working principle of the FP interferometer is shown in Figure 4q. When light enters the FP interferometer, it penetrates the first mirror. The light then interacts with the second mirror, which reflects a portion of the light back toward the first mirror; the rest is transmitted. The reflected light from the second mirror gets reflected by the first mirror and arrives at the second mirror again. This continues, resulting in multiple reflections and interferences between the transmitted and reflected light waves.

The FP interference effect was utilized to realize surface roughness recognition [22]. The structure was shown in Figure 4r. The silver film and the metallic hemispherical nozzle tip worked as the two mirrors in an FP interferometer. However, the light output direction was reversed, compared to Figure 4q. The photonic receptor was on the same side as the light source. The reflected interference light was sorted out from the incident light by the circulator. When the nozzle tip moved on a rough surface (Figure 4s), the spacing between two FP interferometer mirrors kept changing, resulting in continuous variations of reflected light intensity. Then, this ever-changing reflected light intensity data was analyzed by KNN and SVM. The surface roughness was classified into 16 categories, and the best performance was 84.2% using KNN.

- 3.5 Discussion

From the fused optical fiber and machine learning applications, different methods have their advantages and disadvantages. For FBG, its main advantage is that it is sensitive to bending and can be distributed at different positions in series to collect information simultaneously. However, FBG is expensive to manufacture and usually requires a circulator at the light source end, which makes the whole system more complicated. The strength of Specklegram lies in the ability of its images to contain complex and diverse information. Furthermore, because machine learning has far more advantages than other technologies in image processing, the joint application of specklegram and machine learning is worth putting more effort into. Of course, specklegram has its drawbacks. First, a camera must be installed at the receiving end to collect the complete image. Second, recent articles all use the structure of SMF-MMF-SMF, and the connection between optical fibers makes the system construction more complicated. The advantages and disadvantages of light intensity are all attributed to its simplicity. The advantage is that it is simple and cheap without additional requirements at the receiving end. The disadvantage is that its too-simple working principle leads to an inevitably small amount of information it can carry. This shortcoming significantly limits its development ceiling. The advantage of the FP interferometer is its sensitivity to small-amplitude high-frequency vibrations. Like specklegram, since it collects images, it can make full use of the powerful capabilities of machine learning in image processing. Its limitation is that its structure is special and complicated to make, and it needs to add a circulator and a camera at the receiving end, making the system even more complex.

Overall, the advantage of general optical fiber sensors is that they are not affected by the electromagnetic environment, their structures are stable and therefore durable, and the users feel comfortable wearing them because they do not involve electricity. The disadvantage is that they require light sources and receiving devices, affecting the portability and increasing the cost.

An interesting phenomenon about optical fiber is that most of the articles related to FBG and specklegram use MLP and CNN. On the contrary, articles on light intensity and FP interferometer all use traditional machine learning. In this regard, we have the following understandings. Specklegram itself comes in the form of images. FBGs are usually applied in series, forming a two-dimensional dataset, like images. Image-like data characteristics make CNN the best choice for these two applications. While, the predominant use of traditional machine learning in articles involving light-intensity applications is probably due to the reason that light intensity variation was studied relatively early, and the use of traditional machine learning in this area is mature, so it is more feasible to continue using this traditional machine learning instead of deep learning.

3 Electrical Fiber Sensors

Electrical fiber sensors are the other big family of fiber-shaped sensing devices. By working with machine learning, there are four main categories of electrical fiber sensors: piezoresistive, triboelectric, myoelectric, and chip-in-fiber packaging. We first classify electrical fiber sensors based on their working principles. Secondly, unlike optical fiber sensors, which are all millimeter-scale in diameter, electrical fiber sensors have both macroscopic fibers with millimeter-scale diameters and microscopic fibers with micron-scale and even nano-scale. Macroscopic fibers appear in the form of normal fibers, while microscopic fibers are usually sprayed on the base material or dispersed in the base material and finally displayed in the form of a film. Therefore, based on working principle classification, each category is further divided into macroscopic fiber sensors and microscopic fiber sensors.

- 1 Piezoresistive Sensors

Piezoresistive sensors are a type of sensor that utilizes the piezoresistive effect to determine mechanical stress or strain by measuring electrical resistance. The piezoresistive effect refers to the phenomenon where the electrical resistance of certain materials changes when they are subjected to mechanical deformations. Various materials exhibit piezoresistive properties, including silicon, germanium, graphene, carbon nanotubes (CNTs), polyvinylidene fluoride (PVDF), etc. Although their detailed working principles may differ, the general idea is the same, when there is a mechanical deformation, the microstructure of materials changes, resulting in resistance variations [23].

- 1.1 Macroscopical Piezoresistive Fibers

Macroscopic piezoelectric fibers adopt the same structure: a core fiber covered by functional materials. The core in the middle does not need to be conductive, and its main function is to support the shape of the fiber sensor. The functional material of the outer layer is responsible for realizing the electrical function. There are also various processes for making such fibers: dip coating, kneading, electrospinning, etc.

Carbon black (CB) was dip-coated on polyether urethane urea (PUU) to construct a piezoresistive sensor (Fig. 5a) [24]. The fiber core was the polyether urethane urea responsible for flexibility, and the carbon black formed the piezoresistive layer. Each segment of PUU-CB played the role as a strain sensor. This work aimed to realize lower body kinematics monitoring using the fiber sensor mentioned above. Interestingly, this work used a genetic algorithm to determine the best number of sensors deployed and their placements. Then, data from nine sensors were sent to a deep CNN model for joints' angle prediction. The root mean square error (RMSE) and normalized root mean square error (NRMSE) of the test were lower than 6.4∘ and 10%.

On the basis of core fiber and functional materials, an additional layer of ecoflex was added for protection [25, 26]. The piezoresistive layer was made of CNTs, and the core fiber was made of polyurethane (PU) (Fig. 5b) [25]. The PU fiber was put

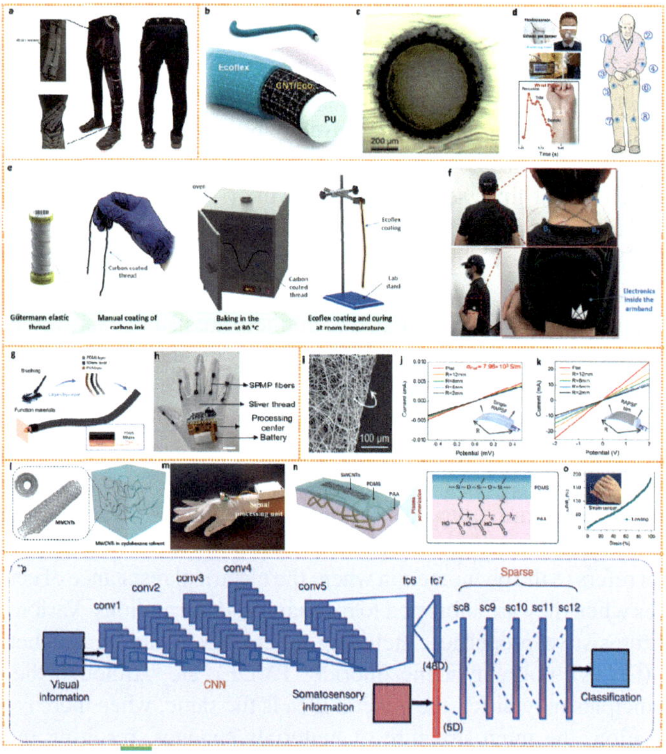

Fig. 5 **a** The fabric-based strain sensors and motion tracker markers mounted on the left leg. **b** Schematic diagram of the fibrous strain sensor, including PU foundation, Eco/CNT conductive layer, and Ecoflex encapsulation layer. **c** Cross-sectional optical microscope image of the fiber sensor. **d** Respiratory monitoring system, wrist pulse monitoring system, and body posture monitoring system. Reproduced with permission. [25] Copyright 2022, Wiley–VCH GmbH. **e** The fabrication process of the strain-sensitive thread includes a manual coating of carbon, baking at 80 °C, and Ecoflex coating. **f** Thread placement on the back of the neck for head motion monitoring. **g** Schematic depicting the fabrication process. The SPMP fiber consists of the spandex core, the PVA layer, the MXene layer, and the PDMS layer (from the inside to the outer side). **h** Picture depicting the hybrid intelligent system involving SPMP fibers, conductive silver threads, microcontroller, and battery. Reproduced with permission. [27] Copyright 2021, Elsevier Ltd. **i** SEM image of RAPSF film exposed to bending. **j** Single RAPSF and **k** RAPSF film with different bending radii on the PET substrates. **l** Illustration of dispersing MWCNTs into cyclohexane solvent through ultrasonic treatment. **m** A smart glove with fiber sensors sewn into each finger and data visualization through Bluetooth. Reproduced with permission. [29] Copyright 2022, Elsevier Ltd. **n** Schematic showing the multilayer stacked structure of the conformable, transparent, and adhesive stretchable strain sensor. SWCNTs, PDMS, and PAA hydrogels function as the sensing layer, the stretchable substrate, and the adhesive layer, respectively. **o** Strain resistance response curves of the stretchable strain sensor under loading. Inset: a strain sensor patched over the knuckle of an index finger. Reproduced with permission. [30] Copyright 2020, Springer Nature Limited. **p** Schematic showing how visual and somatosensory information is processed and fused in the BSV-associated learning architecture. conv., convolutional layer; fc, fully connected neural network; sc, sparsely connected neural network; 48D, 48-dimensional visual vector; 5D, five-dimensional somatosensory vector. blue boxes represent the AlexNet CNN, pretrained using the ImageNet dataset. The five-layer sparse neural network (sc8 to sc12) was used to fuse the SV information

into CNT powder and kneaded for 15–20 min to make the CNTs embedded on the surface of the PU fiber. Then, the fiber was encapsulated by the Ecoflex0050, and the cross-section image was shown in Fig. 5c. This piezoresistive fiber sensor was then used for waist pulse monitoring, respiratory monitoring, and body posture tracking (Fig. 5d). In these experiments, each strain sensor was made of a short segment of PU-CNT-Ecoflex fiber. A deep learning model was used for body posture tracking, and the relative resistance change of the piezoresistive sensor was the input. According to the image, the model used was an MLP, and the average accuracy was 92.7%.

The piezoresistive layer could also be created by coating carbon resistive ink on the surface of Gütermann elastic thread (64% Polyester, 36% Polyurethane) (Fig. 5e) [26]. This sensor was then used for head motion classification. Two fibers were pasted crosswise at the back of the neck (Fig. 5f). Different head motions resulted in various resistance variation pairs, which became the input to machine learning models. 3 KNN variants and 4 SVM variants had been tested on this 9-motion classification task. Accuracies of these models ranged from 89.5% to 92.2%, where linear SVM achieved the best performance.

A four-layer piezoresistive fiber sensor was produced using a so-called 'brushing' technique [27]. There were four layers in total (Fig. 5g): the first core fiber was a spandex yarn, the second adhesive layer was made of polyvinyl alcohol (PVA), the third sensing layer was composed of MXene, and the fourth protective layer was made of polydimethylsiloxane (PDMS). This fiber sensor was used for underwater hand gesture recognition (Fig. 5h). Five waterproof MXene fibers were woven into the glove at the position of five fingers; when a finger moved, the resistance of the corresponding fiber altered. These five sets of resistance variations were sent to an MLP to realize underwater hand gesture recognition, achieving a 98.1% classification accuracy for 20 kinds of hand gestures.

– 1.2 Microscopical Piezoresistive Fibers

Different from macroscopical fiber, microscopical ones usually come in the form of films, but the effective piezoresistive materials inside are still fiber-shaped. For microscopical piezoresistive fibers, there are two types of structures: one is surface film formation, and the other is percolation. The surface film formation gathers a certain amount of piezoresistive materials on the surface of a flexible substrate, forming a sensing layer. While, instead of integrating piezoresistive materials into the substrate, the percolation structure makes the whole membrane a piezoresistive sensor.

A typical example of surface film formation is shown in Fig. 5i–k [28]. It used phase separation and temperature-controlled grain growth strategy to spin an ultra-light piezoresistive material called radial anisotropic porous silver fiber (RAPSF) on the surface of a piece of polyethylene terephthalate (PET). This sensor was then attached to the throat to monitor four types of throat motion: 'Goodbye,' 'Hello,' swallowing, and coughing. The current from this sensor became the input to machine learning models, including KNN, SVM, and XGBoost, where XGBoost achieved the best accuracy of 91.4% for throat motion classification.

Instead of forming a film on top of the substrate, percolation merged piezoresistive nanofibers into the flexible substrate. Multi-wall carbon nanotubes (MWCNTs) could be merged into Ecoflex (Fig. 5l) [29]. MWCNTs were first dispersed into cyclohexane through ultrasonic treatment, and then Ecoflex was added to the prepared CNT/cyclohexane mixture. Next, the ultimate mixture was drawn into fibers using a syringe. The piezoresistive effect relied on the conductive pathways created by direct contact and tunneling effect among interconnected and adjacent MWCNTs. When the fiber was stretched, MWCNTs in the fiber were separated, making the tunneling resistance greater. This piezoresistive fiber was then incorporated into a glove for gesture recognition (Fig. 5m). Each finger had a piece of fiber attached, and finger movements caused resistance changes. These five sets of resistance change information were processed by a CNN model called LeNet-5 for gesture recognition. For a four-class gesture classification task, this model achieved 98.4% accuracy.

A three-layer stretchable sensor consisting of a sensing layer composed of single-wall carbon nanotubes (SWCNTs), a stretchable layer of polydimethylsiloxane (PDMS), and an adhesive layer made of poly (acrylic acid) (PAA) hydrogel was designed (Fig. 5o) [30]. When the film was stretched, conductive pathways between SWCNTs became weaker, resulting in greater resistance (Fig. 5p). This transparent sensor was also utilized for gesture recognition, and five sensors were attached to the knuckles. First, visual pictures of gestures taken by cameras were fed into a CNN model. Second, the output vector of CNN and its corresponding somatosensory data from stretchable sensors were sent to a sparse neural network for data fusion and final gesture recognition (Fig. 5q). With the help of somatosensory data, this approach achieved 100% accuracy for ten-category gesture recognition.

- 2 Triboelectric Nanogenerators

Piezoresistive sensors change resistance passively and do not produce any power, while another important electrical phenomenon, the triboelectric effect, can actively generate power on its own. The triboelectric effect is a phenomenon that occurs when two different materials come into contact and then separate, resulting in the transfer of electric charges between them [31]. The key to the charge transfer is the materials' electron affinity or ionization energy. In other words, some materials prefer positive charges, while others prefer negative ones. When materials with opposite preferences come into contact, one takes positive charges, and the other takes negative charges, resulting in the triboelectric effect. The triboelectric effect is more widely used in energy harvest in the field of wearable technology, but it can also be applied to sensing.

– 2.1 Macroscopical Triboelectric Fibers

As mentioned above, the triboelectric effect requires contacts and fictions between two materials of opposite electron affinities. So, the most straightforward way is to weave the two triboelectric fibers with opposite electron affinities together. Epoxy and polyvinylidene fluoride (PVDF) were woven into a mesh with a copper wire as the conductive core (Fig. 6a, b) [32]. This mesh was then embedded into a mask for respiratory monitoring. As the mask user breathed, airflow stretched the embedded mesh so that epoxy fiber and PVDF fiber contacted with each other, causing the triboelectric effect. The electrical potential created by charge transfer was eventually converted into the current in the copper wire. In this work, breathing was classified into five groups: normal breathing, deep breathing, occasional cough, normal voice, and respiratory failure. Each breathing mode produced a specific current output. This on-mask sensor network provided five channels of current signals to a 1D CNN for the respiration pattern recognition with a classification accuracy of 100%,

Different from weaving conductive wires between triboelectric fibers, embedding silver wires into one of the triboelectric materials was another solution. Nylon, polytetrafluoroethylene, and silver fibers were used to create a self-powered sensing textile for sitting posture monitoring (Fig. 6c, d) [33]. Polytetrafluoroethylene covered the silver wire, and then Ag/polytetrafluoroethylene fibers and nylon fibers were woven into a mesh. Contacts and separations of nylon and polytetrafluoroethylene produced the triboelectric effect, and the resulting current was collected by the silver wire for six-sitting-posture classification. Decision trees, logistic regression, and random forest were the classifiers that had been used. The random forest classifier achieved the best accuracy, 96.6%, higher than the logistic regression (95.5%) and decision tree (94.3%) classifiers.

In addition to weaving two triboelectric materials together, another approach was to use human skin as one of the triboelectric materials. A Cu-silicone fiber was designed for silent speech detection [34]. Copper and silicone did not trigger a triboelectric effect, but when human skin got in touch with silicone, a charge transfer happened, and the electron flow was captured by the copper wire (Fig. 6e). In the application, Cu-silicone fibers were produced and rearranged to form a sensing film which was then integrated into a surgical mask for silent speech detection (Fig. 6f). The movement of the user's face during silent speaking caused membrane deformation and thus changed the device-skin contact area. The triboelectric current was the input to machine learning algorithms. Linear and quadratic discriminant analysis, linear, quadratic, and Gaussian support vector machine, and K-nearest Neighbor had been tried to identify the word silently spoken by the user. Test words included the number "three," the letter "D," and the word "print.", and the Gaussian SVM model gave the highest classification accuracy of 99.2%.

– 2.2 Microscopical Triboelectric Fibers

Individual triboelectric fibers in micro- or nano-scale are usually too small to effectively generate power. In order to facilitate effective power generation and collection, microscopical triboelectric fibers are heavily aggregated and sprayed onto the substrate materials.

CNTs and thermoplastic elastomer (TPE) could be sprayed on the surface of textiles, forming a self-powered conductive superhydrophobic triboelectric textile for gesture recognition [35]. The CNTs and TPE were first mixed in cyclohexane and sprayed on the surface of textiles. The cyclohexane later got volatilized, and the coated textile was then washed with ethanol to form a porous and hydrophobic structure on the surface (Fig. 6g). This textile was then attached to a glove at the joints of ten fingers (Fig. 6h). During gesturing, the current produced by this triboelectric

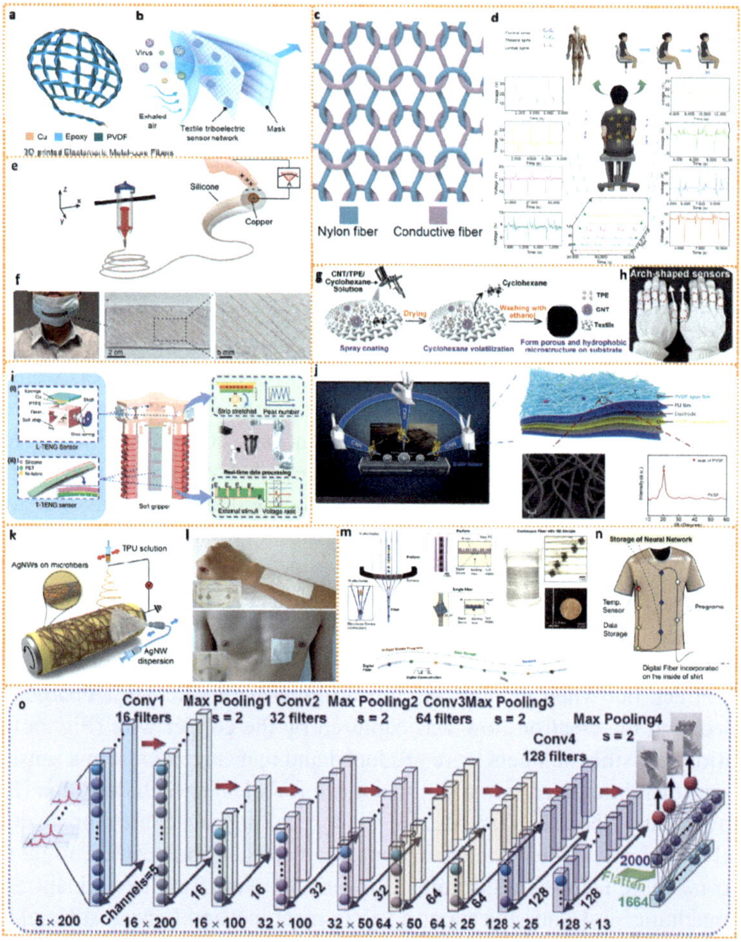

Fig. 6 **a** Schematic showing the structure of the textile sensor. **b** Design of an on-mask sensor network. Reproduced with permission. [32] Copyright 2022, Wiley–VCH GmbH. **c** Knitted loop fiber structure with interlaced double threads. **d** Schematic illustration of signal response to human sitting postures. Reproduced with permission. [33] Copyright 2022, Tsinghua University Press. **e** Concept of 3D printing elastomeric metal-core silicone-Cu fibers. **f** Photograph of a human subject wearing the triboelectric membrane integrated facemask. Reproduced with permission. [3] Copyright 2020, Elsevier Ltd. **g** The fabrication process of the superhydrophobic textile. **h** A glove designed for hand gesture identification. **i** Construction drawing of the low-cost triboelectric nanogenerator (TENG) for soft gripper and its digital twin applications. **j** Concept display of gesture recognition and control through the tactile sensor system. Insets include the schematic diagram of the tactile sensor, the SEM image of electrospinning PVDF, and the XRD of PVDF electrospun film. Reproduced with permission. [37] Copyright 2022, American Chemical Society. **k** Schematic illustration of the experimental setup to generate conductive textiles based on AgNWs and TPU microfibers. **l** Optical images showing epidermal sensing patches attached to the forearm for EMG recording (top) and the chest for ECG recording (bottom). Reproduced with permission. [3, 40] Copyright 2021, American Chemical Society. **m** Fabrication approach of digital fibers. **n** Schematic of the shirt integrated with a digital fiber with sensors, data storage, customizable programs, and a neural network stored within its memory devices. **o** The structure of the CNN model for gesture recognition

textile varied so that machine learning could utilize this information for gesture recognition. The model used was a classical CNN model (Fig. 6o). The voltages of ten sensors were the input, and the gesture classification was the output. In the most complex experiment consisting of 11 gestures, an average accuracy of 95.23% was achieved.

The same way was adopted to construct the triboelectric film, but the carrier was a soft robot gripper instead of a real hand [36]. Two types of triboelectric nanogenerators (TENG) were designed: Length-TENG and Tactile-TENG. They both utilized triboelectric nanofiber nickel-fabrics in design. As shown in Fig. 6ii, the L-TENG used a gear coated with nickel-fabrics as the positive triboelectric component, while the negative triboelectric layer was composed of a polytetrafluoroethylene (PTFE) film. The PTFE film was equipped with a Cu electrode on its backside, which was connected to the external output. The structure of the T-TENG sensor was shown in Fig. 6i (ii), five electrodes made of Ni-fabric were patterned on the polyethylene terephthalate (PET) substrate, and a layer of silicone rubber, as the negative triboelectric material, was coated on the surface of the patch. Signals from L-TENG and T-TENG were then fed into an SVM model for grip object determination. Six objects were involved: apple, cube, long can, orange, short can, and tape, and the average prediction accuracy was 97.083%.

A self-powered tactile sensor was designed for gesture recognition based on PVDF [37]. Interestingly, PVDF was made into not only a triboelectric nanogenerator but also a piezoresistive sensor (Fig. 6j). The top layer, a PVDF electrospun film, and the third layer, an electrode, formed a triboelectric nanogenerator, which provided a power supply function for the sensor. The second and last layers were polyurethane (PU) films, acting as the protective layer. The middle layer, a PVDF film, and the upper and lower electrodes formed a piezoelectric nanogenerator. The piezoelectric PVDF formed a serpentine mesh structure between two electrodes. Whenever the piezoresistive film was subjected to the pressure, the internal crystal structure lost its central symmetry, and the anions and cations that were previously aligned became separated. This separation led to the creation of an electric dipole, generating the piezoelectric effect [38]. The triboelectric nanogenerator acted as a power source, and the current flowed between the triboelectric nanogenerator and piezoelectric sensor varied according to the pressure. In the case of gesture recognition, four tactile sensors were attached to the back of the hand, and four current signals were collected and analyzed by a CNN model. Six gestures were included in this gesture classification task: 'One,' 'Two,' 'Three,' 'Five,' 'Six,' and 'Eight.', and an average accuracy rate of 94.16% was achieved.

- 3 Electromyography Collectors

In addition to triboelectric generators, electromyography is another type of measure that can collect human body information without external power. Electromyography (EMG) records the muscle's reaction or electrical signals in response to a nerve's stimulation [39].

Thermoplastic polyurethane (TPU) microfibers were electrospun onto a textile, and silver nanowires (AgNWs) were simultaneously deposited by passing pressurized AgNW dispersion in isopropyl alcohol through a spray nozzle (Figure 6k) [40]. AgNWs network played the role of a conductive layer. The TPU microfiber scaffold supported the coherent deformation of the AgNWs network and meanwhile provided strong interfacial adhesion to prevent conductive network damage upon stretching. This microtextile exhibited excellent stretchability, conductivity, washability, and breathability. Therefore, it was then used as the electrode of electromyography (EMG) (Figure 6l). The multichannel EMG collected was then fed into a gradient-boosting machine learning algorithm, termed extreme gradient boosting (XGboost), to realize gesture recognition. The final recognition accuracy was not provided.

- 4 Chip in Fiber

With the recent development of functional fibers, instead of involving piezoresistive or triboelectric sensors, pre-fabricated chips could be directly embedded into a single fiber [41]. A fiber of tens of meters containing hundreds of interspersed digital temperature sensors and memory devices was produced through the thermal drawing. The fiber-chip structure was well-thought, and each chip was positioned at a precise angle of 26.56° relative to the fiber axis (Figure 6m). This particular angle enabled the proper alignment of four coplanar and coaxial tungsten wires, allowing them to individually connect to the four distinct power and signal ports on each chip. This digital fiber was then incorporated within a shirt, collecting and storing temperature-time data and achieving real-time inference of wearer activity (sitting, standing, walking, and running) through a CNN (Figure 6n). The accuracy was 96.4% for this four-class human activity recognition.

- 5 Discussion

Regarding the combination of electrical fiber sensors and machine learning, different methods have their own advantages and disadvantages. For example, piezoresistive fibers are usually low-cost and durable due to their stable structures. However, they require external power supplies to realize resistance change measurements and are usually sensitive to the temperature. In contrast to piezoresistive fibers, triboelectric fibers generate electricity by themselves and, therefore, do not require any external power supply. The disadvantage is that their structures are not stable and prone to irreversible changes after use, making their service life shorter. As for myoelectricity, although the electrical signal of the human body is used and no external energy supply is required, the fiber itself only acts as a conductor. Chip-in-fiber is a creative idea. It endows the fiber with the ability of on-site data processing by embedding chips into fibers. However, fibers only play the role of soft shells, which is auxiliary, similar to the fiber in myoelectricity.

The general advantage of electrical fiber is that many materials are available, enabling many functions. The limitation is that they may be affected by the electromagnetic environment, the presence of electricity may cause uncomfortable wearing, and some structures are unstable with limited service life.

Among the articles about electrical fibers, CNN is the most popular model. Others, like MLP, SVM, KNN, and decision trees, have some cases but appear less frequently. The extensive use of CNN is closely related to the extensive application of electrical fibers in mechanical force sensing. Force or deformation information is always described in 2D, close to the form of images, so it is reasonable to see the popularity of CNN here. However, rather than saying that electrical fibers are closely related to CNN, it would be more reasonable to state that force and deformation information are consistent with CNN.

4 Machine Learning in Fiber-Based Wearable Devices

The history of machine learning can be traced back to the mid-twentieth century. At the Dartmouth conference in 1956, the concept of artificial intelligence was proposed, and machine learning became a field of study. Over the following fifty years, many machine algorithms emerged, including multilayer perceptron (MLP), support vector machine (SVM), boosting, and deep neural networks. Although machine learning kept developing in the last century, it was not widely adopted. Until 2012, the event of the ImageNet challenge, a convolutional neural network (CNN) called AlexNet, developed by Alex Krizhevsky, won the competition with unprecedented accuracy of a top-5 error rate of 15.3%. This accuracy surpassed other algorithms at that time by a large margin. Over the next few years, algorithms based on CNN continued to achieve better performance, GoogleNet 6.7%, ResNet 3.57%, and SENet 2.25%. According to human experiments on ImageNet, the error rate of human beings is from 5 to 10%. These data show that machine learning has matched or even surpassed human beings in certain fields. The excellent performance of machine learning in this challenge drives people to apply machine learning in many areas. In various applications, some special abilities of machine learning have been proven: the ability to handle complex and nonlinear relationships, scalability (efficiency in handling large datasets), and the ability to uncover insights and patterns in data, etc. In the field of wearable technology, information from the human body is usually complex, in large volumes, and sometimes obscure. These attributes make machine learning a suitable approach for data analysis in wearable technology.

Machine learning algorithms have many ways to categorize. In this chapter, we roughly divided them into two large groups: conventional machine learning and deep learning. Deep learning is picked out because it leads to the recent rise of machine learning, and it often outperforms other algorithms on complex tasks. However, not all applications require deep learning, and deep learning needs a considerable amount of data and computation budget to support its excellent performance. Hence, many articles still use conventional machine learning algorithms to achieve their goals.

- 1. Conventional Machine Learning Algorithms

– 1. 1 Linear Regression

The first algorithm, which is also the most straightforward, is linear regression [42]. It tries to model the linear relationship between the dependent and independent variables. Figure 7a shows the effect of linear regression intuitively. The following equation represents the primary form of a linear regression model:

$$Y = A_0X_0 + A_1X_1 + A_2X_2 + \cdots + A_nX_n \tag{1}$$

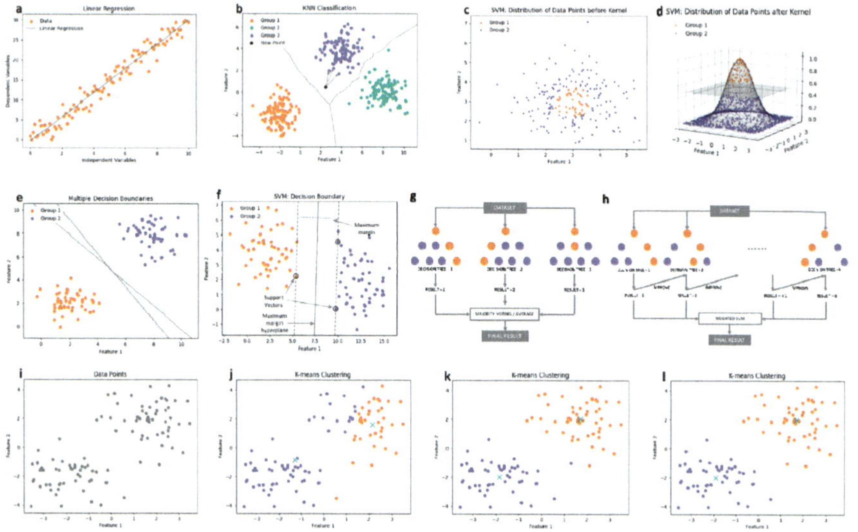

Fig. 7 **a** Illustration of simple linear regression. **b** KNN classification; the new data point is assigned to the group that the majority of K neighbors belong to. **c** SVM kernel trick: before the kernel trick, two groups cannot be separated by a straight line. **d** SVM kernel trick: after the kernel trick, two groups can be easily separated by a plane. **e** SVM maximum margin: without further conditions, there are an infinite number of boundaries that can separate two groups. **f** SVM maximum margin: the hyperplane that maximizes the margin between the support vectors is selected as the boundary. **g** Schematic illustration of random forest. **h** Schematic illustration of XGBoost. **i** Data distribution of two categories before K means clustering. **j** Two centroids are randomly initialized, and data points are assigned accordingly. **k** Two centroids are recalculated, and data points are reassigned. **l** KMC terminates until clusters stop changing or the maximum iteration limit is reached

where Y is called the dependent variable or prediction, $X_0, X_1, \ldots, X_n$ are independent variables or inputs, and $A_0, A_1, \ldots, A_n$ are coefficients or weights of independent variables. The objective is to minimize the error 'e' between the prediction Y and the Ground Truth G by adjusting the coefficients $A_0, A_1, \ldots, A_n$. To facilitate further calculations, the error e_s is expressed as the square of the difference between Y and G.

$$e_s = (Y - G)^2$$

This is the case of one sample. In most cases, multiple samples are provided for training. Error from every case needs to be counted. So, for a group of samples, the error becomes the sum of the squared difference.

$$e = \sum (Y_i - G_i)^2 \tag{2}$$

To achieve error minimization, mathematical optimization techniques have to be used. The most common technique is calculus, and the partial derivatives of the error e with respect to each coefficient are set to zero. Solving these equations yields the coefficient estimates that minimize the difference, e.

$$\frac{\partial e}{\partial A_0} = 0; \frac{\partial e}{\partial A_1} = 0; \ldots; \frac{\partial e}{\partial A_n} = 0 \tag{3}$$

During testing, a piece of information consisting of $A_0, A_1, \ldots, A_n$ is substituted into Eq. 1 to produce a prediction Y.

This is the general process of a linear regression model. It is straightforward, with only linear relationships between predictions and inputs, so it is computationally efficient. However, its drawbacks are also evident: it is sensitive to outliers and cannot deal with nonlinear relationships, hence seldom used for complicated tasks.

– 1.2 K Nearest Neighbors (KNN)

KNN is a non-parametric machine learning algorithm for classification and regression tasks [43]. Classification produces discrete categorical output, while regression produces continuous output. 'Non-parametric' means that KNN does not involve any parameter or coefficient adjustments, thus all the training data and corresponding labels are stored at once. When a piece of test data comes in, the distances, most commonly Euclidean distance, between the new data point and all the data points in the training set are calculated. Then, K, which is set in advance, nearest neighbors are found based on distances. If it is a classification problem, the label of the majority of these K neighbors is the label of the new data point, as shown in Fig. 7b. If it is a regression problem, the average value of K neighbors is the output. KNN is simple and robust to outliers, but it cannot deal with a large dataset, because it needs to store all the training data and calculate the distance between the test data point and every training data point.

– 1.3 Support Vector Machine (SVM)

The idea of SVM is to find an optimal hyperplane in a high-dimensional feature space that separates the data points belonging to different classes with the largest possible margin [44]. This hyperplane is determined by support vectors, the closest data points from each class to the decision boundary. The following content includes the two most important steps to realize SVM.

1. Kernel trick: SVM first uses kernel functions to transform the original input data (Fig. 7c) into a higher-dimensional feature space, where the classes become more separable (Fig. 7d). Here are some commonly used kernels: polynomial kernel $K(x, y) = (\alpha * x^T * y + c)^d$, sigmoid kernel $K(x, y) = tanh(\alpha * x^T * y + c)$, and gaussian radial basis function kernel $K(x, y) = exp(-\gamma * (|x - y|)^2)$. Kernels are chosen according to different needs in various applications.
2. Margin maximization: After kernel transformation, data points become separable. However, as shown in Fig. 7e, multiple valid hyperplanes can be found to separate two classes without constraint; therefore, the criteria for picking the best one is of great significance. SVM proposes to select the hyperplane that maximizes the margin between the support vectors of different classes. The margin is the distance between the hyperplane and the closest data points from each class, as shown in Fig. 7f. This margin maximization problem can be expressed as the following equation.

$$max_{\boldsymbol{w},b} \frac{\gamma(\boldsymbol{w}, b)}{||\boldsymbol{w}||^2} s.t. \forall i y_i(\boldsymbol{w}^T \boldsymbol{x}_i + b) \geq 0$$

$$where \gamma(\boldsymbol{w}, b) = min_{\boldsymbol{x}_i \in D}(|\boldsymbol{w}^T \boldsymbol{x}_i + b|)$$

$max_{\boldsymbol{w},b} \frac{\gamma(\boldsymbol{w},b)}{||\boldsymbol{w}||^2}$ maximizes the margin, $\forall i y_i(\boldsymbol{w}^T \boldsymbol{x}_i + b) \geq 0$ represents the separating hyperplane, and $\boldsymbol{w}$ is the normal vector of the hyperplane. This expression can eventually be simplified to

$$min_{\boldsymbol{w},b} \boldsymbol{w}^T \boldsymbol{w} s.t. \forall i y_i(\boldsymbol{w}^T \boldsymbol{x}_i + b) \geq 1$$

This is a quadratic optimization problem that can be solved by a dedicated quadratic optimization solver, and the solution is unique. During testing, the attributes $X_1, X_2, \ldots, X_n$ of the new data point are substituted into the expression of hyperplane $\boldsymbol{w}^T \boldsymbol{x}_i + b$. If the result is greater than 0, it is classified into class A. Otherwise, it is classified into class B.

– 1.4 Random Forest

Random forest is an ensemble learning method that combines multiple decision trees to make predictions (Fig. 7g) [45]. Since its structure involves many decision trees, it is called a forest. A decision tree is a hierarchical structure comprising a series of decisions and their possible consequences. The top node is the root node representing the initial decision, the internal nodes represent decision points, and the leaf nodes represent the outcomes or class labels. The algorithm selects a splitting characteristic at each internal node to divide the data into subsets. For a complex event, tons of properties are available to become splitting characteristics for the current node. Therefore, various property selection strategies lead to totally different decision trees. Not only does the selection strategy influence tree growth, but different training datasets also lead to different tree structures due to different data distributions. Individual decision trees may have bias, variance, and sometimes overfitting problems. To overcome these limitations of a single decision tree, random forest trains multiple decision trees with random training subsets using different property selection strategies and, in the end, takes the average or highest vote as the output, avoiding overfitting and improving the resistance to the noise.

– 1.5 XGBoost

Similar to the random forest, XGBoost is also based on decision trees [46]. Random Forest uses multiple decision trees in parallel, but XGBoost generates decision trees in sequence. Its structure is shown in Fig. 7h. The XGBoost realization is described as the following steps.

1. Make an initial prediction and calculate residuals.
2. Build an XGBoost decision tree based on similarity and gain.
3. Prune the tree based on gain and threshold gamma.
4. Calculate output values of leaves.
5. Make new predictions based on output values.
6. Calculate new residuals based on new predictions.
7. Repeat step 2 to step 6 until the residual is small enough or the maximum iteration is reached.
8. The final prediction is the weighted sum of trees that emerges from the previous steps. The formula is shown as follows:

$$T_0 + \varepsilon_1 T_1 + \varepsilon_2 T_2 + \varepsilon_3 T_3 + \cdots + \varepsilon_n T_n$$

where T's are XGBoost trees emerge during adjusting.

– 1.6 K Means Clustering (KMC)

Unlike all the supervised learning algorithms above, K means clustering [47] is an unsupervised learning algorithm, meaning that KMC groups data points according to features, ignoring the ground truth. A KMC process (Fig. 7i–l) contains five steps:

1. Decide the number of clusters 'K.'
2. Initialize cluster centers or centroids randomly.
3. Assign data points to their closest centroid.
4. Recalculate the clusters' centers by taking the mean of data points in the same group.
5. Repeat steps 3 and 4 until clusters stop changing or the algorithm reaches the maximum iteration limit.

- 2. Deep Learning Algorithms

Deep learning is the main reason why machine learning has rapidly developed in the past decade. It excels in processing large volumes of data and captures intricate patterns and relationships that may be difficult for other algorithms to uncover. Deep learning can also extract features automatically without manual engineering and keep learning with growing datasets. In a word, deep learning stands out thanks to its excellent performance and special abilities not found in other algorithms. Such a powerful tool certainly has applications in wearable fiber sensors.

– 2.1 Multilayer Perceptron (MLP) / Backpropagate Neural Network

Multilayer perceptron or backpropagate neural network is the cornerstone of deep learning [48]. It consists of an input layer, one or more hidden layers, and an output layer. Each layer contains multiple neurons or nodes, and all layers are fully connected, as shown in Fig. 8a. In other words, each neuron in one layer is connected to every neuron in the next layer. Each neuron in an MLP performs a weighted sum of its inputs, followed by an activation function of the sum, and produces an output. This operation is called a neuron operation. The activation function introduces non-linearity into the network, enabling it to model complex relationships, for example, the sigmoid function, hyperbolic tangent function, and rectified linear unit (ReLU). Take the top node of the first hidden layer as an example, $y = \sigma(w_{11}x_1 + w_{21}x_2 + w_{31}x_3 + w_{41}x_4)$, where w_{ab} represents the weight from the ath node in the previous layer to the bth node in the current layer, and σ is the sigmoid function defined as $\sigma(x) = 1/(1 + exp(-x))$.

There are two steps in each epoch or loop of training: forward propagation and backpropagation. The forward propagation lets the data flow from the input layer through every hidden layer, experiencing a series of neuron operations mentioned above, and comes out at the output layer. After forward propagation, backpropagation is executed to update all the weights between nodes, and can be considered to be

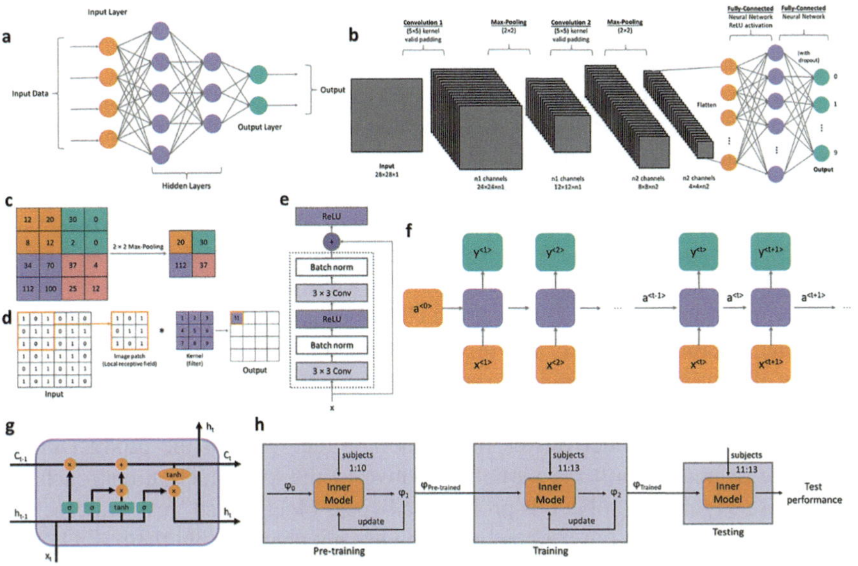

Fig. 8 **a** Schematic illustration of a multilayer perceptron. **b** Schematic illustration of a convolutional neural network. **c** An example of max-pooling. **d** An example of convolution. **e** Residual block: layers in the dashed box do not change, and the bypass channel on the right is the key to the ResNet. **f** Schematic illustration of a recurrent neural network. **g** Schematic illustration of a hidden block in LSTM. **h** Schematic illustration of transfer learning

the essence of MLP. It first compares the outputs with ground truths, calculates the loss function, and takes the derivative of the error with respect to the weights and biases in the network. This derivative, also called gradient, indicates the direction and magnitude of the adjustments that nodes need to minimize the error. The gradients are passed to nodes layer by layer from the output layer back to the input layer. Because the parameter passing direction is opposite to the forward propagation, this weights update process is called backpropagation. Upon the completion of a pair of forward and backward propagations, an epoch or loop of training has been completed. The whole training process is just a repetition of forward and backward propagation.

– 2.2 Convolutional Neural Network (CNN) and ResNet

The convolutional Neural Network is a deep neural network that is particularly effective in analyzing visual data such as images [49]. Figure 8b illustrates the structure of a basic CNN. The last section of CNN is a layer called the fully connected layer, which is exactly the same as a multilayer perceptron. The key improvement of CNN over MLP is its image preprocessing, consisting of max pooling (Fig. 8c) and convolution (Fig. 8d). The convolutional layer performs the convolution operation by applying learnable filters (kernels) to the input data. Each filter is a small matrix of weights that slides across the input, computing dot products with the receptive field at each spatial location. The purpose of convolution is to extract specific patterns in the input. The max-pooling downsamples the output of the convolution layer by preserving only the max value of each image patch. Max-pooling makes the network more robust to input variations while preserving the most important information. Max-pooling also reduces the number of parameters, which would be useful when the dataset becomes larger, and the network becomes deeper. In conclusion, CNN comprises convolution layers, max-pooling layers, and a fully connected layer. The training procedure is the same as MLP, starting from a forward propagation, followed by a backpropagation adjusting the weights of neuron operations in the fully connected layer and kernels in convolution layers. Such a closed loop forms a training epoch of CNN.

When the number of hidden layers of CNN becomes larger, that is a very deep CNN, and a problem called 'vanishing gradient' occurs. As mentioned in CNN, weights of neuron operations and kernels are updated during backpropagation, and the essence of backpropagation is derivative. As the path of backpropagation gets longer, the number of derivatives increases, and the value of the product of the derivative decreases until the partial derivative of the loss function approaches a value close to zero. To solve this problem, a bypass block called 'residual block' is proposed. As shown in Fig. 8e, the backbone has not changed, but there is a bypass channel that allows 'x' to skip the following convolution layers and get summed up with the output directly. This method decreases the depth of backpropagation to some extent and allows information from further ancestors to be preserved, relieving the vanishing gradient problem. The CNN model with such residual blocks is called ResNet [50].

– 2.3 Recurrent Neural Network (RNN) and Long Short Term Memory (LSTM)

CNN is good at image processing but does not deal with sequential or time series information, for example, which image comes first. In order to process sequential data, a new type of neural network is introduced: Recursive Neural Network (RNN) [51]. The general structure of an RNN is shown in Fig. 8f. The blocks at the bottom are sequential inputs x, purple blocks in the middle represent the hidden layer that contains neuron operations, including weights and activation, and a's between blocks are information passed down from precious data. Each purple block has two outputs: 'y' the prediction is shown in green blocks on the top, and 'a' the information to be passed down. For each timestep t, the activation $a^{<t>}$ and the output $y^{<t>}$ are expressed as follows:

$$a^{<t>} = g_1\left(W_{aa}a^{<t-1>} + W_{ax}x^{<t>} + b_a\right) and y^{<t>} = g_2(W_{ya}a^{<t>} + b_y)$$

Same as MLP, the prediction y is compared with the ground truth to produce the error and accordingly update weights in blue boxes by backpropagation through time (BPTT). Such a forward and backward propagation similar to MLP is an epoch of RNN training.

Although RNN includes sequential information, each hidden block only inherits information from its direct ancestor and cannot receive information from the ancestor two blocks away. In other words, standard RNNs do not describe long-term dependencies. To deal with this issue, long short term memory (LSTM) is proposed [52].

The structure of LSTM is shown in Fig. 8g. This block is just the purple block in RNN. The critical component in LSTM is the cell state represented by C, which is the convey belt of memory through the network. The following steps happen sequentially in one neuron at one timestep.

1. Decide what information to forget from the cell state. The f in the figure is called the forget gate.

$$f_t = \sigma(W_f \cdot [h_{t-1}, x_t] + b_f)$$

2. Decide what new information is to be added to the cell state. The i is called the input gate.

$$i_t = \sigma(W_i \cdot [h_{t-1}, x_t] + b_i)$$

$$\tilde{C}_t = \tanh(W_C \cdot [h_{t-1}, x_t] + b_C)$$

3. Update cell state.

$$C_t = f_t * C_{t-1} + i_t * \tilde{C}_t$$

4. Finally, the output is generated.

$$o_t = \sigma(W_o \cdot [h_{t-1}, x_t] + b_o)$$

$$h_t = o_t * \tanh(C_t)$$

- 2.4 Transfer Learning

In real-life applications of machine learning, training data can be hard or costly to collect. One solution to this problem is transfer learning [53]. Transfer learning trains the model, which is called the pre-trained model, using more easily obtained data from relative domains, and then the pre-trained model is fine-tuned using the data from the exact target domain. This method greatly reduces the difficulty of training data collection. As demonstrated in Fig. 8h, the respiratory flow information from 13 subjects has been collected [7]. In the beginning, the parameters of LSTM are randomly initialized. Pre-training takes the data from 10 subjects to pre-train the LSTM model. Then, in the training phase or the so-called fine-tuning, instead of random initialization, parameters of pre-trained LSTM are preserved as the starting parameters. Data from subjects 11–13 are divided into training datasets and testing datasets. Training data is used to fine-tune the LSTM in the training phase, while testing data is used to test the transfer-trained LSTM model in the testing phase.

- 3 Discussion

As mentioned above, there are many structures and algorithms, whether it is traditional machine learning or deep learning. Most machine learning algorithms have good adaptability and can be adjusted and applied to various scenarios. However, apparent preferences exist, and some models are naturally more suitable for some applications due to their inherent structure. The Convolutional Neural Network (CNN) is specially designed for image processing. As images are essentially matrices of pixels, CNN innately has strong adaptability to simultaneous multi-channel signal processing. Hence, CNN well suits tasks about force and shape changes, such as item determination based on force distribution on gloves, where the data arrangement is akin to an image. The Recurrent Neural Network (RNN), on the other hand, takes the time influence, or in another word, context, into consideration. Signals at different times are sequentially put together for learning. Therefore, this model is widely used in natural language processing. In the field of wearable devices, when it comes to data collected over a period of time and with a continuous relationship, for example, respiratory disease diagnosis based on breathing patterns, RNN usually produces good results.

Although machine learning is very powerful, it has limitations in application. The first thing to bear the brunt is the demand for data. The accuracy and representativeness of training data will significantly affect the effectiveness of the model. Recent powerful models require high-quality and often large quantities of data. Therefore, the quality and quantity of collected data are a major bottleneck limiting machine learning applications on wearable devices. The second point is the demand for computing resources. Machine learning, especially DL, involves a lot of computations. The current solution is to upload data to the server for calculation. On the one hand, this will require data bandwidth. On the other hand, since both data uploading and downloading take time, it will cause delays. Therefore, two research directions can be considered to address these limitations. One is to achieve better sensors to improve the quality and quantity of data. The other is edge computing, which reduces the amount of upstream and downstream data and server computing pressure.

Data quality and quantity limitations are common problems in machine learning applications. In the field of fiber-based wearable devices, there is a special problem. Most models mentioned in this review were created before 2018. As the updating speed of machine learning algorithms is extremely fast, it is necessary to apply new models more boldly in fiber-based wearables. Personal speculation is that, on the one hand, the data collection capabilities of fiber sensors are not as mature as general-purpose sensors, resulting in insufficient data volume to meet the needs of new models. On the other hand, the existing fiber sensor application scenarios are relatively simple, resulting in simple tasks, and there is no need for using complex models. In addition to boldly engaging new models, whether it is data volume or task complexity, their solution seems to lie in the performance of fiber sensors, such as data accuracy, stability, diversity, information richness, etc. This leads to our next section about fiber sensors.

5 Conclusion

Fiber sensors are a feasible solution to replace hard electronic sensors in wearable devices, and machine learning combined with fiber sensors can realize smart clothing. This paper has reviewed fiber sensors and classified them into optical and electrical fiber sensors according to their working principles. Optical fiber sensor includes FBG, interference image, intensity, and FP interferometer. Electrical fiber sensors include piezoelectric, triboelectric, EMG, and chip-in-fiber modes. Then, they are further divided into macroscopic and microscopic fibers according to their scales. Next, the machine learning algorithms used in these articles are also summarized. The traditional machine learning algorithms include linear regression, K nearest neighbors, support vector machine, and random forest. Deep learning includes multilayer perceptron (MLP), convolutional neural network (CNN), recurrent neural network (RNN), and transfer learning.

Although many articles have applied machine learning to fiber-based wearable devices, we believe that the application is still in an early stage. It has the following characteristics, which are also problems and directions for future improvements. First of all, most applications only focus on mechanical force and deformation related topics, like glove-based gesture recognition. However, the information that wearable devices can collect is far from limited to mechanical force. The intensity and color of outside light, the temperature and humidity of the environment, the roughness of the skin, and the composition of sweat are all very meaningful information that can be collected. The second point is that most works only collect one signal at a time. The collection of only one signal greatly limits the information capacity, and the simplicity of the information also limits the space for machine learning. The third problem is that the machine learning algorithms used, especially the deep learning methods, are more conventional. New and powerful models have not been actively engaged yet, and most of the works simply adapt the machine learning algorithm from libraries directly without fine-tuning the algorithm according to the application scenario. Fourthly, which is also one of the reasons for the third problem, is that current tasks are relatively simple. The dataset size is small, making it hard to fully explore the powerful functions of deep learning.

We propose to solve or alleviate the problems mentioned above from two perspectives: hardware and software. From the hardware side, we need to improve the performance of fiber sensors. Here, we already have a promising solution, which is the multifunctional fiber [54]. It has been applied in the field of neurobiology, collecting electrical, chemical, and mechanical signals simultaneously [55]. Moreover, the multifunctional fiber is naturally compatible with a field of machine learning called data fusion, which combines multiple data sources to produce more consistent, accurate, and useful information [56, 57]. If the multifunctional fiber is used properly, it can well solve the oversimplicity problem of data collected and alleviate the embarrassment of insufficient data volume to a certain extent. As for software, machine learning has developed rapidly in recent years. New and powerful structures are emerging one after another. We should promptly try cutting-edge machine learning algorithms in the field of fiber sensor information processing, such as reinforcement learning, generative adversarial network (GAN), self-supervised learning, and attention mechanism (GPT). GAN can be used to supplement data [58], self-supervised learning can reduce the cost of labeling [59], and attention mechanisms can enhance prediction capabilities [60]. These are directions that we believe are promising for future research. With the help of fiber sensors and the latest machine-learning algorithms, wearable devices will become more comfortable, efficient, and intelligent.

References

1. Hill KO, Meltz G. Fiber Bragg grating technology fundamentals and overview. J Light Technol. 1997;15(8):1263–76. https://doi.org/10.1109/50.618320.
2. Zhuang Y, et al. A fiber-optic sensor-embedded and machine learning assisted smart helmet for multi-variable blunt force impact sensing in real time. Biosensors. 2022;12(12):1159. https://doi.org/10.3390/bios12121159.
3. Li T, et al. Bioinspired stretchable fiber-based sensor toward intelligent human-machine interactions. ACS Appl Mater Interfaces. 2022;14(19):22666–77. https://doi.org/10.1021/acsami.2c05823.
4. Massari L, et al. A machine-learning-based approach to solve both contact location and force in soft material tactile sensors. Soft Robot. 2020;7(4):409–20. https://doi.org/10.1089/soro.2018.0172.
5. Massari L, et al. Functional mimicry of Ruffini receptors with fibre Bragg gratings and deep neural networks enables a bio-inspired large-area tactile-sensitive skin. Nat Mach Intell. 2022;4(5):425–35. https://doi.org/10.1038/s42256-022-00487-3.
6. Lun TLT, Wang K, Ho JDL, Lee K-H, Sze KY, Kwok K-W. Real-time surface shape sensing for soft and flexible structures using fiber bragg gratings. IEEE Robot Autom Lett. 2019;4(2):1454–61. https://doi.org/10.1109/LRA.2019.2893036.
7. Filosa M, et al. A meta-learning algorithm for respiratory flow prediction from FBG-based wearables in unrestrained conditions. Artif Intell Med. 2022;130: 102328. https://doi.org/10.1016/j.artmed.2022.102328.
8. Saccomandi P, et al. Feedforward neural network for force coding of an MRI-compatible tactile sensor array based on fiber bragg grating. J Sens. 2015;2015:1–9. https://doi.org/10.1155/2015/367194.
9. Ha XT, et al. Shape sensing of flexible robots based on deep learning. IEEE Trans Robot. 2023;39(2):1580–93. https://doi.org/10.1109/TRO.2022.3221368.
10. Sefati S, Gao C, Iordachita I, Taylor RH, Armand M. Data-driven shape sensing of a surgical continuum manipulator using an uncalibrated fiber bragg grating sensor. IEEE Sens J. 2021;21(3):3066–76. https://doi.org/10.1109/JSEN.2020.3028208.
11. Li T, et al. Flexible optical fiber-based smart textile sensor for human-machine interaction. IEEE Sens J. 2022;22(20):19336–45. https://doi.org/10.1109/JSEN.2022.3201580.
12. Leal-Junior AG, Frizera A, Marques C, Pontes MJ. Optical fiber specklegram sensors for mechanical measurements: a review. IEEE Sens J. 2020;20(2):569–76. https://doi.org/10.1109/JSEN.2019.2944906.

13. Cuevas AR, Fontana M, Rodriguez-Cobo L, Lomer M, Lopez-Higuera JM. Machine learning for turning optical fiber specklegram sensor into a spatially-resolved sensing system. Proof of concept. J Light Technol. 2018;36(17):3733–8. https://doi.org/10.1109/JLT.2018.2850801.
14. Sun K, Ding Z, Zhang Z. Fiber directional position sensor based on multimode interference imaging and machine learning. Appl Opt. 2020;59(19):5745. https://doi.org/10.1364/AO.394280.
15. Ding Z, Zhang Z. 2D tactile sensor based on multimode interference and deep learning. Opt Laser Technol. 2021;136: 106760. https://doi.org/10.1016/j.optlastec.2020.106760.
16. Ma S, et al. Optical micro/nano fibers enabled smart textiles for human-machine interface. Adv Fiber Mater. 2022;4(5):1108–17. https://doi.org/10.1007/s42765-022-00163-6.
17. Wang S, et al. Optical-nanofiber-enabled gesture-recognition wristband for human-machine interaction with the assistance of machine learning. Adv Intell Syst. 2023;5(7):2200412. https://doi.org/10.1002/aisy.202200412.
18. Avellar L, Stefano Filho C, Delgado G, Frizera A, Rocon E, Leal-Junior A. AI-enabled photonic smart garment for movement analysis. Sci Rep 2022;12(1):4067, https://doi.org/10.1038/s41598-022-08048-9.
19. Leal-Junior A, Avellar L, Frizera A, Marques C. Smart textiles for multimodal wearable sensing using highly stretchable multiplexed optical fiber system. Sci Rep. 2020;10(1):13867. https://doi.org/10.1038/s41598-020-70880-8.
20. Huang H, Lin J, Wu L, Fang B, Wen Z, Sun F. Machine learning-based multi-modal information perception for soft robotic hands. Tsinghua Sci Technol. 2020;25(2):255–69. https://doi.org/10.26599/TST.2019.9010009.
21. Vaughan M, The Fabry-Perot interferometer: history, theory, practice and applications. Routledge (2017).
22. Keser S, Hayber ŞE. Fiber optic tactile sensor for surface roughness recognition by machine learning algorithms. Sens Actuators Phys. 2021;332: 113071. https://doi.org/10.1016/j.sna.2021.113071.
23. Fiorillo AS, Critello CD, Pullano SA. Theory, technology and applications of piezoresistive sensors: a review. Sens Actuators Phys. 2018;281:156–75. https://doi.org/10.1016/j.sna.2018.07.006.
24. Gholami M, Rezaei A, Cuthbert TJ, Napier C, Menon C. Lower body kinematics monitoring in running using fabric-based wearable sensors and deep convolutional neural networks. Sensors. 2019;19(23):5325. https://doi.org/10.3390/s19235325.
25. Gao J, et al. Ultra-robust and extensible fibrous mechanical sensors for wearable smart healthcare. Adv Mater. 2022;34(20):2107511. https://doi.org/10.1002/adma.202107511.
26. Jiang Y, Sadeqi A, Miller EL, Sonkusale S. Head motion classification using thread-based sensor and machine learning algorithm. Sci Rep. 2021;11(1):2646. https://doi.org/10.1038/s41598-021-81284-7.
27. Duan S, et al. Machine-learned, waterproof MXene fiber-based glove platform for underwater interactivities. Nano Energy. 2022;91: 106650. https://doi.org/10.1016/j.nanoen.2021.106650.
28. Lin S, et al. An ultralight, flexible, and biocompatible all-fiber motion sensor for artificial intelligence wearable electronics. Npj Flex Electron. 2022;6(1):27. https://doi.org/10.1038/s41528-022-00158-8.
29. Liu Z, et al. Flexible strain sensing percolation networks towards complicated wearable microclimate and multi-direction mechanical inputs. Nano Energy. 2022;99: 107444. https://doi.org/10.1016/j.nanoen.2022.107444.
30. Wang M, et al. Gesture recognition using a bioinspired learning architecture that integrates visual data with somatosensory data from stretchable sensors. Nat Electron. 2020;3(9):563–70. https://doi.org/10.1038/s41928-020-0422-z.
31. Wu C, Wang AC, Ding W, Guo H, Wang ZL. Triboelectric nanogenerator: a foundation of the energy for the new era. Adv Energy Mater. 2019;9(1):1802906. https://doi.org/10.1002/aenm.201802906.
32. Fang Y, et al. A deep-learning-assisted on-mask sensor network for adaptive respiratory monitoring. Adv Mater. 2022;34(24):2200252. https://doi.org/10.1002/adma.202200252.

33. Jiang Y, et al. Knitted self-powered sensing textiles for machine learning-assisted sitting posture monitoring and correction. Nano Res. 2022;15(9):8389–97. https://doi.org/10.1007/s12274-022-4409-0.
34. Tong Y, Feng Z, Kim J, Robertson JL, Jia X, Johnson BN. 3D printed stretchable triboelectric nanogenerator fibers and devices. Nano Energy. 2020;75; 104973. https://doi.org/10.1016/j.nanoen.2020.104973.
35. Wen F, et al. Machine learning glove using self-powered conductive superhydrophobic triboelectric textile for gesture recognition in VR/AR applications. Adv Sci. 2020;7(14):2000261. https://doi.org/10.1002/advs.202000261.
36. Jin T, et al. Triboelectric nanogenerator sensors for soft robotics aiming at digital twin applications. Nat Commun. 2020;11(1):5381. https://doi.org/10.1038/s41467-020-19059-3.
37. Yang J, et al. Self-powered tactile sensor for gesture recognition using deep learning algorithms. ACS Appl Mater Interfaces. 2022;14(22):25629–37. https://doi.org/10.1021/acsami.2c01730.
38. Wang ZL, Song J. Piezoelectric nanogenerators based on zinc oxide nanowire arrays. Science. 2006;312(5771):242–6. https://doi.org/10.1126/science.1124005.
39. Mills KR. The basics of electromyography. J Neurol Neurosurg Psychiatry 2005;76(suppl_2):ii32–ii35, https://doi.org/10.1136/jnnp.2005.069211.
40. Zhao H, et al. Ultrastretchable and washable conductive microtextiles by coassembly of silver nanowires and elastomeric microfibers for epidermal human-machine interfaces. ACS Mater Lett. 2021;3(7):912–20. https://doi.org/10.1021/acsmaterialslett.1c00128.
41. Loke G, et al. Digital electronics in fibres enable fabric-based machine-learning inference. Nat Commun. 2021;12(1):3317. https://doi.org/10.1038/s41467-021-23628-5.
42. Montgomery DC, Peck EA, Vining GG. Introduction to linear regression analysis. Wiley (2021).
43. Peterson LE. K-nearest neighbor. Scholarpedia. 2009;4(2):1883. https://doi.org/10.4249/scholarpedia.1883.
44. S. Suthaharan, Machine learning models and algorithms for big data classification: thinking with examples for effective learning. In: Integrated Series in Information Systems, vol 36. Boston, MA: Springer US (2016). https://doi.org/10.1007/978-1-4899-7641-3.
45. Biau G, Scornet E. A random forest guided tour. TEST. 2016;25(2):197–227. https://doi.org/10.1007/s11749-016-0481-7.
46. Chen T, Guestrin C. Xgboost: A scalable tree boosting system. In: Proceedings of the 22nd acm sigkdd international conference on knowledge discovery and data mining, pp 785–794 (2016).
47. Hartigan JA, Wong MA. Algorithm AS 136: A k-means clustering algorithm. J R Stat Soc Ser C Appl Stat. 1979;28(1):100–108.
48. Noriega L. Multilayer perceptron tutorial.
49. O'Shea K, Nash R. An introduction to convolutional neural networks. arXiv, Dec. 02, 2015. Accessed Jun. 20, 2023. [Online]. Available: http://arxiv.org/abs/1511.08458.
50. He K, Zhang X, Ren S, Sun J. Deep residual learning for image recognition. arXiv (2015). Accessed: Jun. 24, 2023. [Online]. Available: http://arxiv.org/abs/1512.03385.
51. Grossberg S. Recurrent neural networks. Scholarpedia. 2013;8(2):1888. https://doi.org/10.4249/scholarpedia.1888.
52. Staudemeyer RC, Morris ER. Understanding LSTM—a tutorial into long short-term memory recurrent neural networks. arXiv, Sep. 12, 2019. Accessed: Jun. 20, 2023. [Online]. Available: http://arxiv.org/abs/1909.09586.
53. Weiss K, Khoshgoftaar TM, Wang D. A survey of transfer learning. J Big Data. 2016;3(1):9. https://doi.org/10.1186/s40537-016-0043-6.
54. Shen Y et al. Thermally drawn multifunctional fibers: toward the next generation of information technology. InfoMat 2022;4(7), https://doi.org/10.1002/inf2.12318.
55. Canales A, et al. Multifunctional fibers for simultaneous optical, electrical and chemical interrogation of neural circuits in vivo. Nat Biotechnol. 2015;33(3):277–84. https://doi.org/10.1038/nbt.3093.

56. Khaleghi B, Khamis A, Karray FO, Razavi SN. Multisensor data fusion: a review of the state-of-the-art. Inf Fusion. 2013;14(1):28–44. https://doi.org/10.1016/j.inffus.2011.08.001.
57. Meng T, Jing X, Yan Z, Pedrycz W. A survey on machine learning for data fusion. Inf Fusion. 2020;57:115–29. https://doi.org/10.1016/j.inffus.2019.12.001.
58. Aggarwal A, Mittal M, Battineni G. Generative adversarial network: an overview of theory and applications. Int J Inf Manag Data Insights. 2021;1(1): 100004. https://doi.org/10.1016/j.jjimei.2020.100004.
59. Jaiswal A, Babu AR, Zadeh MZ, Banerjee D, Makedon F. A survey on contrastive self-supervised learning. Technologies. 2020;9(1):2. https://doi.org/10.3390/technologies9010002.
60. Niu Z, Zhong G, Yu H. A review on the attention mechanism of deep learning. Neurocomputing. 2021;452:48–62. https://doi.org/10.1016/j.neucom.2021.03.091.

Index

L. Wei (ed.), *Advanced Optical and Optoelectronic Fibers*, Advances in Optics and Optoelectronics, https://doi.org/10.1007/978-981-97-6218-7